Uncovering Student Ideas in Astronomy

45 NEW Formative Assessment Probes

Uncovering Student Ideas in Astronomy

45 NEW Formative Assessment Probes

By Page Keeley
and Cary Sneider

NSTApress
National Science Teachers Association
Arlington, Virginia

Claire Reinburg, Director
Jennifer Horak, Managing Editor
Andrew Cooke, Senior Editor
Wendy Rubin, Associate Editor
Agnes Bannigan, Associate Editor
Amy America, Book Acquisitions Coordinator

ART AND DESIGN
Will Thomas Jr., Director
Cover, Inside Design, and Illustrations by Linda Olliver

PRINTING AND PRODUCTION
Catherine Lorrain, Director
Nguyet Tran, Assistant Production Manager

NATIONAL SCIENCE TEACHERS ASSOCIATION
Francis Q. Eberle, PhD, Executive Director
David Beacom, Publisher

1840 Wilson Blvd., Arlington, VA 22201
www.nsta.org/store
For customer service inquiries, please call 800-277-5300.

15 14 13 12 4 3 2 1

ISBN 978-1-936137-38-1
eISBN 978-1-936959-82-2

Cataloging-in-Publication Data is available from the Library of Congress.

Contents

Section 5. Stars, Galaxies, and the Universe

Dedication

This book is dedicated to an astronomy educator, artist, dedicated NSTA workshop presenter, and all-around brilliant and wonderful person—Donna Young. Thank you, Donna, for your tireless efforts in bringing the wonders of the universe to teachers everywhere and supporting the next generation of astronomy educators.

Foreword

There is nothing like astronomy to pull the stuff out of man.
His stupid dreams and red-rooster importance: let him count the star-swirls.
—From the poem "Star-Swirls" by Robinson Jeffers

As the old adage has it: *You can't know where you're going until you know where you've been.* And it's every bit as true in science education as it is in other parts of life. In this intriguing book, Page Keeley and Cary Sneider, two experienced and talented educators, set out to help teachers figure out where their students have been in their thinking about our planet and its place in the universe, so that your teaching and their learning of astronomical ideas can then be as effective as possible.

After many decades of research about learning, today we know that students are not simply blank vessels waiting to be filled with our brilliantly taught lessons. They're more like hot dogs you buy from a street vendor—already full, although not necessarily with material you would approve of. The more you know about what your students' heads are filled with, the better you will be at helping them.

Students of each generation construct their understanding of the world from the best resources they have available—their common sense, their daily experience, the media they are exposed to, conversations with family and friends. Today, the internet plays a much bigger role than it did for the previous generation, but the process by which students absorb a wild mixture of information and approaches in the course of their lives has not changed. And we should always appreciate that students have tried to make the best synthesis of all these influences that they could. After all, their mental constructs—whether scientifically correct or not—show the scientist's urge to make sense of the world and put disparate phenomena into a reasonable context.

This book consists of a series of small but carefully thought-out probes into student thinking about astronomy. Just like a doctor's diagnostic tool provides one chemical or physical indicator of our health, each of Keeley and Sneider's probes measures one or two ideas that lets you know how much surgical repair (if any) might be needed to fix up your students' astronomical ideas.

Astronomy is a great portal to help students solidify their thinking about science and the scientific process. Surveys show that for younger children, astronomy and dinosaurs continue to be the most fascinating entrées into science (and you need only look at popular films for youngsters to see this confirmed). But the study of astronomy has something for students of all ages, from simple observations of everyday sky phenomena to considerations of the origins of all heavier elements in the nuclear crucibles of the stars. The sheer beauty of the images being taken by the Hubble Space Telescope and giant mirrors on Earth is a wonderful entrée for students who are visual learners. And then there is the possibility that one day a large chunk of rock or ice from space may again threaten life on Earth as it did 65 million years ago (in the time of the dinosaurs). If you want to capture the imagination of students for science, astronomical lessons and activities may be your strong suit.

Although I have organized and taught summer workshops for teachers and informal educators for three decades, most of my teaching has been and continues to be for non-

Foreword

science college students. In my evening classes, I have students ranging in age from 16 to 75 and so I have seen a nice cross section of the adult American public over the years. I can attest to the fact that these older students are often grappling with the very same misconceptions that are highlighted in this book.

Many of my college students firmly believe that the seasons are caused by the Earth being farther away from the Sun in winter, that the changing appearance (phases) of the Moon is the result of the Earth's shadow falling on the Moon, and that shooting stars are stars that die. Some students have also become convinced by internet and TV conspiracy theorists that NASA never landed on the Moon, that UFOs are alien spaceships whose landings on Earth are hushed up by the government, and that the world will end on the winter solstice in 2012 (as supposedly predicted by the ancient Mayans).

In my classes, I do activities that allow students to confront and lay open the preconceptions that they have on many astronomical subjects. Only after we have faced the existence of these ideas can we then do activities to move the students toward finding a new view of astronomical phenomena, based on scientific experiments and observations. Former students who return to speak with me after some time has passed frequently cite these "mind-set"-changing sessions as the highlights of what they remember from my classes.

It is therefore from direct personal experience that I can commend the techniques in this book to your attention and wish you as much pleasure in teaching the wonders of astronomy as I have had over the years.

—Andrew Fraknoi
Chair, Department of Astronomy, Foothill College, Los Altos Hills, California
Former Executive Director, Astronomical Society of the Pacific, San Francisco, California

In addition to the positions listed above, Andrew Fraknoi was founding co-editor of the journal Astronomy Education Review, *and he founded and directed Project ASTRO, a national program that trains and brings volunteer astronomers into fourth- through ninth-grade classrooms. Fraknoi is the lead author of* Voyages Through the Universe, *one of the leading college textbooks for introductory astronomy, and he wrote a colorful children's book called* Disney's Wonderful World of Space. *He was selected California Professor of the Year in 2007 by the Carnegie Endowment for Higher Education and won the Andrew Gemant Award from the American Institute of Physics for his work in making connections between physics and culture. Fraknoi was recently named an honorary member of the Royal Astronomical Society of Canada, a distinction shared by such notables as Stephen Hawking and Harlow Shapley. The International Astronomical Union named Asteroid 4859 Asteroid Fraknoi to honor his contributions to science education and outreach (but he hastens to add that it is a very boring asteroid that is in no danger of hitting the Earth!).*

Preface

When I heard the learn'd astronomer,
When the proofs, the figures, were ranged in columns before me,
When I was shown the charts and the diagrams, to add, divide, and measure them,
When I sitting, heard the astronomer where he lectured with much
applause in the lecture-room,
How soon unaccountable I became tired and sick,
Till rising and gliding out I wander'd off by myself,
In the mystical moist night-air, and from time to time,
Look'd up in perfect silence at the stars.
—Walt Whitman, *Leaves of Grass*

In the classic poem "When I Heard the Learn'd Astronomer," Walt Whitman contrasts the dry and dull delivery of a lecture on the universe with the awe-inspiring experience of going outside and looking up at a star-filled night sky. One can draw a parallel to similar experiences students have in school—being passively filled with facts about astronomy from reading textbooks and listening to teachers talk. Wouldn't it be more interesting if students could contemplate the wonders of the universe through questions that spark interest and curiosity and lead them to observe or model astronomical phenomena firsthand? This book is the first in the *Uncovering Student Ideas in Science* series to focus on astronomy. Coauthored with Dr. Cary Sneider, an esteemed "learn'd astronomer," educator, and researcher, this book aims to not only inform teaching but also inspire students to want to know more about the wonders of the Earth, Moon, Sun, our solar system, and the universe beyond.

Series Overview: *Uncovering Student Ideas in Science*

The *Uncovering Student Ideas in Science* series provides science educators with unique sets of versatile formative assessment probes for use in K–12 education, preservice teacher programs, and professional development. This popular series of award-winning books began with *Uncovering Student Ideas in Science, Vol. 1: 25 Formative Assessment Probes* (Keeley, Eberle, and Farrin 2005). That book contains 25 K–12 probes in life, Earth, space, and physical sciences. The introductory chapter provides an overview of formative assessment and a description of the standards- and research-based probe development process. Between that 2005 book and this new volume on astronomy, five other books have been published in the series:

- *Uncovering Student Ideas in Science, Vol. 2: 25 More Formative Assessment Probes* (Keeley, Eberle, and Tugel 2007): This book contains 25 K–12 probes in life, Earth, space, and physical sciences. The introductory chapter describes the link between formative assessment and instruction and suggests ways to embed the probes into your teaching. This book won the Distinguished Achievement Award given by the Association of Educational Publishers (AEP).

- *Uncovering Student Ideas in Science, Vol. 3: Another 25 Formative Assessment Probes* (Keeley, Eberle, and Dorsey 2008): This book contains 22 K–12 probes in life, Earth, space, and physical sciences and 3 nature of science probes. The introductory chapter describes how to use the probes for teacher learning either through individual study or in a variety of professional development formats.
- *Uncovering Student Ideas in Science, Vol. 4: 25 New Formative Assessment Probes* (Keeley and Tugel 2009): This book contains 23 K–12 probes in life, Earth, space, and physical sciences and 2 unifying concepts probes, one on models and one on systems. The introductory chapter describes the link between formative and summative assessment, including why an investment in formative assessment throughout an instructional cycle can improve student performance on summative assessments. This book was also a winner of the Distinguished Achievement Award and a finalist in the Golden Lamp Award given by the AEP.
- *Uncovering Student Ideas in Physical Science, Vol. 1: 45 New Force and Motion Assessment Probes* (Keeley and Harrington 2010). Coauthored with physicist and physics education researcher Dr. Rand Harrington, this book contains 45 K–12 probes organized in the categories of Describing Position and Motion; Forces and Newton's Laws; and Mass, Weight, Gravity, and Other Topics. The introductory chapter provides an overview of the research, teaching, and student ideas related to force and motion, including advice for implementing the probes. Other volumes planned for this physical science series will cover electricity and magnetism, sound and light, and matter and energy. This book was also a winner of the AEP's Distinguished Achievement Award.
- *Uncovering Student Ideas in Life Science, Vol. 1: 25 New Formative Assessment Probes* (Keeley 2011b): This book includes 25 biology/life science probes organized in the categories of Life and Its Diversity; Structure and Function; Life Processes and Needs of Living Things; Ecosystems and Adaptation; Reproduction, Life Cycles, and Heredity; and Human Biology. The introductory chapter describes how formative assessment probes are used in a life science context. Two more books are planned for this life science series.

Collectively, this new astronomy book and the six volumes that preceded it now provide educators with 215 assessment probes. In addition, the introductory chapters of all the books in the series can significantly expand teachers' assessment literacy, understanding of the ideas students bring to their learning, key ideas that make up fundamental topics in science, effective teaching and learning strategies, and access to other related resources.

About the Probes in This Book

The probes in this book, as well as others in the series, are diagnostic assessments designed to uncover students' thinking. Once students' ideas are revealed, the teacher uses the information to make decisions about instruction that will help students give up their misconceptions or incompletely formed ideas in favor of the scientific view or a more coherent and connected understanding of the content. It is this use of the probe to inform instruction that transitions it from being a diagnostic assessment to a formative assessment. If one uncovered students' preconceptions but did nothing with the information, then the probe would be merely diagnostic. It is the formative use of the probes that make them powerful resources for promoting learning and improving teaching.

Although these probes are printed on paper and are often used as written formative assessments, their real value comes through providing opportunities for students to discuss their ideas and formulate scientific arguments to support their thinking while evaluating and challenging the ideas of their peers. Whether used at the beginning of a unit as an elicitation or at checkpoints during various stages of instruction, the probes are designed to draw out ideas commonly held by students. Beyond having students merely select an answer, the real value of the probes emerges when students are asked to explain their thinking. It is this part of the probe that provides many insights into students' ways of thinking and reasoning that draw on their everyday experiences outside of school as well as ideas "learned" in school that failed to come together in a coherent, connected, or accurate way.

All of the probes in this series have common features: a title designed to spark interest; a prompt set in an engaging, familiar context; a set of answer choices to choose from; and an open-ended section where students are asked to provide an explanation. This format is designed for the purpose of making the probe accessible to all students, providing an answer that is most likely to match ideas students have and eliminate the blank stares or "I don't know" responses that often accompany open response formats. The explanation section is also used to promote student thinking and provide opportunities for students to construct explanations to support their thinking. Constructing explanations and defending scientific ideas through talk and argument are key science practices included in *A Framework for K–12 Science Education*, which will inform the Next Generation Science Standards tentatively due to be released in late 2012 (NRC 2011).

In addition to these common features of the probes, there are a variety of formats used throughout the *Uncovering Student Ideas in Science* series to draw out students' ideas in different and sometimes novel ways. These formats and the process for developing these types of probes are described in Page Keeley's National Science Foundation–funded Curriculum Topic Study (CTS) materials (Keeley 2005; Mundry, Keeley, and Landel 2009), and the formats are summarized below:

- *Justified List Probes.* These probes provide a list of items which may or may not match a particular concept or statement in the probe's prompt. Students are asked to choose the items on the list they believe match the concept or statement provided and explain their rule or reason for choosing the items. An example of a justified list probe is Probe 16, "Seeing the Moon."
- *Friendly Talk Probes.* These probes involve a group of people talking about their different ideas. Students choose the name of the person they agree with the most and explain why. The format of these probes make them much more engaging than a traditional multiple-choice question where students choose A, B, C, or D. The statements (answer choices) are written to simulate a conversation between people. There is also a degree of safety in choosing the answer in that students worry less about whether they are wrong because they didn't say it, "so and so" did. An example of a friendly talk probe is Probe 30, "Is It a Planet or a Star?"
- *Thought Experiment Probes.* These probes are set in the context of an imaginary scenario that would be impossible to produce in the real world. Students have to use their imagination to visualize the scenario

and apply their knowledge of science to predict a result and explain their reasoning behind their prediction. An example of a thought experiment probe is Probe 3, "Falling Through the Earth."

- *Word Use Probes.* These probes reveal how students confuse the use of everyday words with scientific words or confuse similar-sounding words. An example of a word use probe is Probe 5, "The Two Rs."
- *Opposing Views Probes.* These probes are similar to friendly talk probes except they involve two people with opposite views. Students select the person whose view most matches their own and explain why. An example of an opposing views probe is Probe 27, "Is the Moon Falling?"
- *Follow the Dialogue Probes.* These probes include back-and-forth dialogue between two people. Students follow the dialogue and defend or critique each person's reasoning. An example of a follow the dialogue probe is Probe 42, "Seeing Into the Past."
- *Familiar Phenomenon Probes.* These probes are based on a phenomenon students are familiar with or may have seen or experienced in real life or vicariously. The probe is designed to reveal their ideas about the explanatory nature of the phenomenon. An example of a familiar phenomenon probe is Probe 34, "Shooting Star."
- *Representation Analysis Probes.* These probes provide students with a choice of representations that visually model a concept, process, event, or object. Students choose the representation that best matches their thinking visually and explain why they chose it. An example of a representation analysis probe is Probe 29, "How Do Planets Orbit the Sun?"

Regardless of what type of format is used, all the probes are designed to engage students in surfacing their astronomy ideas and sharing their thinking with other students as well as the teacher. It is for this reason that you are strongly encouraged not to give students the answer to the probe immediately after responding. If you do, the thinking is over! Provide students with time to think the question through, muddle about in reconsidering their ideas as the discussion yields new evidence, and engage in firsthand observations or modeling to discover the answers for themselves. Merely telling students the answer does little to change deeply rooted conceptions. Students need to be vested in the process of realizing for themselves when their ideas no longer work for them. This often happens through rich classroom discourse that encourages and supports the skills of evidence-based argumentation. Eventually students will come to learn the right answer, whether it is through their own process of constructing new knowledge or they are eventually presented with the scientific explanation. Hanging out in uncertainty for a while is not the same as hanging out in uncertainty indefinitely; there is no harm in holding answers back until students are ready to accept the scientific explanation and discard ideas that no longer provide explanatory power for them.

How the Probes Are Arranged in This Book

The probes are arranged in five sections. Section 1 presents ideas related to the nature of our planet, such as the spherical Earth concept and Earth's gravity. Section 2 presents ideas related to the Sun-Earth system, such as the seasons and the Sun's changing path across the daytime sky. Section 3 addresses the Earth-Sun-Moon system, with an emphasis on modeling the Moon. Section 4 goes beyond Earth's moon to explore the dynamic solar system. Finally, Section 5 takes us beyond the solar system to explore ideas related to stars, galaxies, and the universe.

Each section begins with a concept matrix that identifies the major concepts addressed by each probe and the grade span most appropriate for using the probe. Following the matrix is an overview of teaching and learning considerations for the concepts in that section. Next, there is a list of the CTS guides that are often used to generate the probes by linking key ideas in national standards to research-identified misconceptions. CTS guides can also be used to further examine the curricular topic (Keeley 2005). More information about the CTS process and tools is available at *www.curriculumtopicstudy.org.*

The last part of the introductory material in each section is a list of related resources from NSTA and other sources. These highly recommended resources can be used to further develop teachers' content knowledge or provide suggestions for instructional activities and opportunities to further students' learning.

Teacher Notes

Each probe is accompanied by a set of Teacher Notes. These notes contain valuable information about content, teaching, and learning and should be read before using a probe. The sections of the Teacher Notes are described below.

Purpose

This section describes the purpose of the probe—that is, why you would want to use it with your students or the teachers you work with. It begins by describing the overall concept elicited by the probe, followed by the specific idea the probe targets. Before using a probe, you should be clear about what the probe can reveal. Taking time to read the purpose will help you decide if the probe fits your intended use.

Related Concepts

Each probe is designed to target one or more related concepts that often cut across grade spans. These concepts are also included on the matrix charts. A single concept may be addressed by multiple probes as indicated on the concept matrix. You may find it useful to use a cluster of probes to target a concept or specific ideas within a concept. For example, there are several probes that relate to Earth's orbit around the Sun.

Explanation

A brief scientific explanation accompanies each probe and clarifies the scientific content that underlies the probe. The explanations are designed to help you identify what the "best" or most scientifically acceptable answers are (sometimes there is an "it depends" answer) as well as clarify any misunderstandings you might have about the content.

The explanations are not intended to provide detailed background knowledge about the content. In writing these explanations we were careful not to make them so technical that only someone with an astronomy background would understand them, since our intended audience includes teachers who have little or no astronomy or physics background. We provide the information a novice teacher would need to understand the content he or she teaches, but we also try not to oversimplify. If you have a need for additional information about the content, refer to the resources list in each section.

Administering the Probe

In this section, we suggest ways to administer the probe to students, including appropriate grade spans and, in some cases, modifications to make a probe intended for one grade span useful for another. The suggested grade span is intended to be a suggestion only. Your decision about whether or not to use a probe depends on why you are using it and your students' readiness. Do you want to know about the ideas your students are expected to learn in your grade-level standards? Are you interested in how preconceived ideas develop and change across multiple grade levels in your school whether or not they are formally taught? Are you interested in whether students achieved a scientific understanding of previous grade-level ideas before you introduce higher level concepts? Do you have students who are ready for advanced concepts? We recommend that you weigh the suggested grade levels against the knowledge you have of your own students, your school's curriculum, and your state standards.

Related Ideas in the National Standards

This section lists the learning goals stated in the two national documents generally considered the "national standards": *Benchmarks for Science Literacy* (AAAS 1993, 2009) and *National Science Education Standards* (NRC 1996). The learning goals from these two documents are quoted here because almost all states' standards are based on them. Also, since the probes are not designed as summative assessments, the listed learning goals in this section are not intended to be considered alignments but rather ideas that are related in some way to the probe. When the ideas elicited by a probe appear to be a strong match (aligned) with a national standard's learning goal, these matches are indicated by a star (★) symbol. You may find this information useful in using probes with lessons and instructional materials that are strongly aligned to your state and local standards and specific grade level.

Sometimes you will notice that an elementary learning goal is included with middle and high school probes. This is because it is useful to see the related idea that the probe may build on from a previous grade span. Likewise, sometimes a high school learning goal is included even though the probe is designated for grades K–8. This is because it is useful to consider the next level of sophistication that students will encounter in a coherent vertical sequence of learning.

At the time this book was written, the National Research Council had just come out with *A Framework for K–12 Science Education: Practices, Crosscutting Concepts, and Core Ideas* (NRC 2011). This conceptual framework is being used to guide the development of the Next Generation Science Standards that are anticipated to be released in late 2012. Since the state-level implementation phase will most likely not phase in until 2013, and the previous standards still are relevant in states, we chose to keep the Benchmarks and the National Science Education Standards in this volume. However, as we developed the probes, we did notice that the core ideas described in the Framework match well with the probes in this book. Once the Next Generation Science Standards are released, there will be a crosswalk for this book (as well as all the previous books in the *Uncovering Student Ideas* series) that describes the link between each probe and the new standards. That crosswalk will be available on the Uncovering Student Ideas website in 2013: *www.uncoveringstudentideas.org*.

Related Research

Each probe is informed by research on students' astronomy misconceptions, including research conducted by coauthor Dr. Cary Sneider. The research summaries can help you better understand the design and intent of the probe and the kinds of thinking your students are likely to reveal when they respond to the probe. Some of the research studies we cite describe studies that have been conducted in past decades, and they have involved children in both the United States and other countries. Most of the results of these studies are considered timeless and universal. Whether students develop their ideas in the United States or in other countries, research indicates that many of these astronomy ideas are pervasive regardless of geographic boundaries and societal and cultural influences. Teachers who use the probes are encouraged to conduct their own classroom action research with the probes. An example of using probes for action research is described in the *Science and Children* article "Formative Assessment Probes: Is It Living?" (Keeley 2011a). Consider a similar action research approach with these astronomy probes.

Suggestions for Instruction and Assessment

A probe remains simply diagnostic, not formative, unless you use the information about students' ideas to inform instruction. After analyzing your students' responses, the most important step is to decide on the student interventions and instructional paths that would work best in your particular context, based on your students' thinking. We have included suggestions gathered from the wisdom of teachers, the knowledge base on effective science teaching, research, and our own collective experiences working with students and teachers. These are not extensive lists of detailed instructional strategies but rather brief suggestions that may help you plan or modify your curriculum or instruction to help students learn ideas that they may be struggling with. It may be as simple as realizing that you need to provide an opportunity for students to directly observe the phenomenon, or there may be a modeling activity you could use with your students (e.g., to demonstrate the positional relationship of the Moon to the Earth and the Sun when explaining what causes the phases of the Moon). Learning is a very complex process, and most likely no single suggestion will help all students learn. But that is what formative assessment encourages—thinking carefully about the variety of instructional strategies and experiences needed to help students learn important ideas in astronomy. As you become more familiar with the ideas your students have and the multifaceted factors that may have contributed to their misunderstandings, you will identify additional strategies that you can use to teach for understanding.

References

References are provided for the standards, research findings, and some of the instructional suggestions cited in the Teacher Notes. You might want to read the full research summary or access a copy of the research paper or resource cited in the Related Research section of the teacher notes.

Going Formative

Before you begin to use the probes in this book, recall that a probe is not formative unless you use the information from the probe to modify, adapt, or change your instruction so that students have data-informed opportunities to learn the important concepts in astronomy. As a companion to this book, NSTA has copublished *Science Formative Assessment: 75 Practical Strategies for Linking Assessment, Instruction, and Learning* (Keeley 2008). That companion book includes strategies to use with the probes to facilitate elicitation of students' ideas, support metacognition, spark inquiry, encourage discussion and argumentation, monitor progress, give and obtain feedback, improve quality of questions and responses, and promote self-assessment and reflection. In addition, please visit the Uncovering Student Ideas website (*www.uncoveringstudentideas.org*) for additional formative assessment resources and strategies shared by users of the probes. And please look for "Uncovering Student Ideas in Science" sessions offered by NSTA Press at NSTA's area and national conferences and web seminars. We hope to see you there and learn about your (and your students') successes in using these 45 astronomy probes!

Preface

References

American Association for the Advancement of Science (AAAS). 1993. *Benchmarks for science literacy.* New York: Oxford University Press.

American Association for the Advancement of Science (AAAS). 2009. Benchmarks for science literacy online. *www.project2061.org/publications/bsl/online*

Keeley, P. 2005. *Science curriculum topic study: Bridging the gap between standards and practice.* Thousand Oaks, CA: Corwin Press and Arlington, VA: NSTA Press.

Keeley, P. 2008. *Science formative assessment: 75 practical strategies for linking assessment, instruction, and learning.* Thousand Oaks, CA: Corwin Press and Arlington, VA: NSTA Press.

Keeley, P. 2011a. Formative assessment probes: Is it living? *Science and Children* 48 (8): 24–26.

Keeley, P. 2011b. *Uncovering student ideas in life science, vol. 1: 25 new formative assessment probes.* Arlington, VA: NSTA Press.

Keeley, P., F. Eberle, and C. Dorsey. 2008. *Uncovering student ideas in science, vol. 3: Another 25 formative assessment probes.* Arlington, VA: NSTA Press.

Keeley, P., F. Eberle, and L. Farrin. 2005. *Uncovering student ideas in science, vol. 1: 25 formative assessment probes.* Arlington, VA: NSTA Press.

Keeley, P., F. Eberle, and J. Tugel. 2007. *Uncovering student ideas in science, vol. 2: 25 more formative assessment probes.* Arlington, VA: NSTA Press.

Keeley, P., and R. Harrington. 2010. *Uncovering student ideas in physical science, vol. 1: 45 new force and motion assessment probes.* Arlington, VA: NSTA Press.

Keeley, P., and J. Tugel. 2009. *Uncovering student ideas in science, vol. 4: 25 new formative assessment probes.* Arlington, VA: NSTA Press.

Mundry, S., P. Keeley, and C. J. Landel. 2009. *A leader's guide to science curriculum topic study.* Thousand Oaks, CA: Corwin Press.

National Research Council (NRC). 1996. *National science education standards.* Washington, DC: National Academies Press.

National Research Council (NRC). 2011. *A framework for K–12 science education: Practices, crosscutting concepts, and core ideas.* Washington DC: National Academies Press.

Whitman, W. 1965. *Leaves of grass.* New York: New York University Press.

Acknowledgments

We would like to thank all the teachers and science coordinators who tried out the probes with their students, gave us feedback on them, or contributed ideas for further probe development. We especially would like to thank the following educators who carefully reviewed this book, and in many cases pilot tested the probes with their students: Susan Duncan, Beaverton, Oregon; Andrew Fraknoi, San Francisco, California; Nicole Holden, Beaverton, Oregon; Neil Kavanagh, Portland, Oregon; Jaylee Scott, Mesa, Arizona; Haley Candista, Dallas, Texas; and Taryn Devereaux, York, Maine.

We are grateful to our excellent reviewers for providing us with very useful feedback to improve the original manuscript of this book: Ted Willard, Timothy Slater, and Llama Maynard. We especially want to thank Andrew Fraknoi for taking the time to write such a thoughtful and inspirational foreword.

We also appreciate the assistance over the years of the following collaborators and fellow researchers who contributed so much to the research and practice of astronomy education: Jane Abel, Paul Ammon, Varda Bar, William Brewer, Al Byers, Cathy Clemens, Supriya Chakrabarti, David Cudaback, John Dobson, Alan Friedman, Alan Gould, Claudine Kavanagh, Anne Kennedy, Gary Kratzer, Will Kyselka, William Liller, Marcia Linn, Ramon Lopez, Larry Lowery, Gerry Mallon, Cherilynn Morrow, Yossi Nussbaum, Steven Pulos, Philip Sadler, Dennis Schatz, Dwight Sieggreen, Elizabeth Stage, William Waller, and David White.

And finally, we can't say enough about the outstanding support from NSTA Press staff in helping us bring this book to fruition.

About the Authors

Page Keeley is the senior science program director at the Maine Mathematics and Science Alliance (MMSA), where she has worked since 1996. She has directed projects in the areas of leadership, professional development, linking standards and research on learning, formative assessment, and mentoring and coaching, and she consults with school districts and organizations nationally. She has authored 12 books and has been the principal investigator on three National Science Foundation grants: the Northern New England Co-Mentoring Network; Curriculum Topic Study: A Systematic Approach to Utilizing National Standards and Cognitive Research; and PRISMS: Phenomena and Representations for Instruction of Science in Middle School. Most recently she has been consulting with school districts, Math-Science Partnership projects, and organizations throughout the United States on building teachers' capacity to use diagnostic and formative assessment.

Page taught middle and high school science for 15 years and used formative assessment strategies and probes long before there was a name attached to them. Many of the strategies in her books come from her experiences as a science teacher. Page was an active teacher leader at the state and national level. She received the Presidential Award for Excellence in Secondary Science Teaching in 1992 and the Milken National Distinguished Educator Award in 1993, and she was the AT&T Maine Governor's Fellow for Technology in 1994. She has served as an adjunct instructor at the University of Maine, is a Cohort 1 Fellow in the National Academy for Science and Mathematics Education Leadership, and serves on several national advisory boards.

Prior to teaching, she was a research assistant in immunology at the Jackson Laboratory of Mammalian Genetics in Bar Harbor, Maine. She received her BS in life sciences from the University of New Hampshire and her MEd in secondary science education from the University of Maine. Page served as the 63rd President of the National Science Teachers Association. In 2009 she received the National Staff Development Council's Susan Loucks-Horsley Award for her contributions to science education leadership and professional development. Page is a frequent invited speaker at national conferences and led the People to People Citizen Ambassador Programs Science Education delegation to South Africa in 2009, China in 2010, and India in 2011.

About the Authors

Cary Sneider is an associate research professor at Portland State University in Portland, Oregon, where he teaches research methods in a master's degree program for prospective teachers and leads a National Science Foundation–supported project to bridge the gap between high school and college physics. His research interests have focused on helping students unravel their misconceptions in science and on new ways to link science centers and schools to promote student inquiry. He recently served as design lead for technology and engineering on the National Research Council's effort to develop *A Framework for K–12 Science Education: Practices, Crosscutting Concepts, and Core Ideas.* Cary has been involved in the development of framework documents for the National Assessment of Educational Progress (NAEP) in Science and in Technology and Engineering Literacy, and he is currently a member of the National Assessment Governing Board. He is also a leader of the writing team at Achieve, Inc., which is managing the development of the Next Generation Science Standards for the states.

From 1997 to 2007 Cary was vice president for educator programs at the Museum of Science in Boston, where he led development of a high school engineering curriculum, Engineering the Future: Science, Technology, and the Design Process. Prior to that he served as director of astronomy and physics education at the University of California Lawrence Hall of Science. He has taught science at the middle and high school levels in Maine, California, Costa Rica, and Micronesia. In 1997 he received the Distinguished Informal Science Education award from the National Science Teachers Association, and in 2003 he was named national associate of the National Academy of Sciences.

Over his career Cary has directed more than 20 federal and state grant projects, mostly involving curriculum development and teacher education. He earned a bachelor of arts in astronomy from Harvard College and a teaching credential and doctorate from the University of California at Berkeley. He attributes his lifelong passion for astronomy to clear skies in southern Maine, where he spent his formative years, and to his parents and sister Sharon, who put up with the noise and messiness of telescope building.

Introduction

"I don't get it," he said, passing me in the hall one day. Pointing out the window he continued, "Look, there's the Moon. I can see the curved shadow of the Earth on the Moon. But the Sun is still up, so the Earth's shadow must be behind us somewhere. How can Earth's shadow fall on the Moon in the daytime?"

My friend had graduated from an Ivy League school. He had always done well in school and he loved science. As director of a major science education program he certainly had a good formal background in science. But that day I realized he did not understand the modern scientific explanation for Moon phases.

To his credit, at that moment my friend realized that he'd had it wrong all these years. It is likely that he first learned about Moon phases in elementary school, when his spatial reasoning skills had not yet developed. And since he never studied Moon phases again, he never had an opportunity to learn the real reason for phases.

A great many people of all ages make this same mistake. They learn about Moon phases and eclipses, but what really sticks is the explanation for an eclipse of the Moon, also called a lunar eclipse (derived from the Latin word for moon, *luna*). A lunar eclipse occurs when the Moon passes through the Earth's shadow. A lunar eclipse happens about twice a year. Since you have to be on the side of Earth facing the Moon when it is in Earth's shadow, chances are about 50-50 that you'll get to see it. So at any one location it is common to see a lunar eclipse about once a year.

Once we both recognized why he was confused, it was easy to help my friend understand why Moon phases occur. I took a hard-boiled egg out of my lunch bag and we walked outside. I handed him the egg and asked him to hold it up next to the Moon in the sky and pay attention to which part is lighted and which part is in shadow. He figured it out himself. Here's more or less what he said:

"Oh! I see now. The egg and the Moon are both lit up by the Sun. I knew that before. But the shadow on the egg is the same as the shadow on the Moon—and the lighted part is the same, too. In fact, I'm looking at a 'crescent egg'! The dark part is the shadow of the egg on itself, just like with the Moon—the dark part of the Moon is the shadow of the Moon on itself."

For good measure I took a moment to explain to my friend why he had been confused. It's easier to learn the explanation for a lunar eclipse than for Moon phases, and if it is presented too early and then never revisited it's not surprising that someone would remember just one explanation and apply it to both phases and eclipses—even though the explanation for Moon phases is different. Today, thanks to science education research, the concept of Moon phases is more appropriately placed at the upper elementary or middle school level, where many students still find it challenging.

—Cary Sneider, 2011

Why Are Probes Useful for Teaching Astronomy?

The 45 astronomy probes in this book can help you provide the kind of experience in your classroom that Cary and his friend experienced in the exchange described above. Each probe presents a situation that will engage your students' interests while assessing their current level of understanding (or misunderstanding). Although it may not always be easy to help your students untangle mistaken ideas, knowing what your students' preconceptions are is an excellent start.

One way to think about what these probes can do that other assessments often fail to do is to reveal your students' mental models of the world. We all have many such models and we use them all the time. For example, envision in your "mind's eye" the house or apartment building where you live. Imagine you are standing across the street from the front door. What color is the building? How many windows does it have? Is there a mailbox in view? If the mailbox is not in front of the building, can you envision where it is? Now imagine walking across the street, picking up the mail, and entering the building. "Walk around" inside, looking into the different rooms, to become aware of just how detailed this mental model happens to be.

Your mental model of the place you live does much more than provide mild entertainment when reading about science teaching—it provides a map of the world that you follow when you pick up the real mail, walk into your actual house, put away the groceries in the correct cupboards, and find your glasses in the place where you usually leave them. Mental models are such an important part of our lives that we rarely think of them; but without them we could not function in the world. They also color new information that we receive.

Stella Vosniadou and William Brewer (1992, 1994) conducted a series of influential studies on children's mental models in astronomy. They interviewed first-, third-, and fifth-grade students, and found that the children's understanding of the day-night cycle depended on their mental models of the Earth. For example, students whose mental model of the Earth was a flat surface with an absolute "down" in space explained that the Sun would literally go "up" in the daytime and "down" at night. Older students, who held a more advanced spherical Earth concept, understood that the Earth is a ball in space. Those students held different misconceptions. For example, some thought that day and night were caused when the Sun went around the Earth once a day.

Without understanding your students' mental models about the Earth in space, it will be very difficult to help them understand anything that you may want to teach them about astronomy.

Mental Models and the Evolution of Astronomy

Astronomy is sometimes called the "mother of all sciences" because it was the first field to which modern scientific thinking was applied. Although the term *scientist* had yet to be invented, the term might well be applied to Aristotle, the ancient Greek philosopher and teacher of Alexander the Great. Aristotle was born approximately 2,500 years ago in Stagira in what is today northern Greece. He wrote about his ideas on astronomy in a book called *On the Heavens* (see Guthrie 1939 for one of several English translations).

Aristotle's mental model of the Earth in space was remarkably modern in some ways. He described the Earth as a huge sphere (giving proper credit to philosophers of earlier periods) and as evidence of Earth's shape he correctly referred to the appearance of Earth's shadow on the Moon during a lunar eclipse. He noted that as travelers journeyed far to the north or south they would see different stars gradually come into view.

Aristotle's mental model was not entirely modern. He imagined that people could live in many different places on Earth without falling off because in his view Earth was located in the center of the universe. Everything heavy would naturally fall to the center of the universe, while things that were light, such as fire, would go away from the center of the Earth. To explain day and night, he posited that the Sun circled the Earth once a day. Aristotle's writings survived the Middle Ages, and his mental model of the Earth and Sun was taught in the best schools for thousands of years, until it was finally challenged by Copernicus in the 16th century.

Copernicus agreed with Aristotle that the Earth is a sphere. However, he disagreed that the Earth is immovable in the center of the universe. Instead, he proposed that the Sun is immovable, in the center of the universe, and the Earth orbited around it once a year. In his mental model—which is closer to our own—day and night were caused by the Earth spinning once a day as it orbited the Sun once per year.

Since Copernicus worked a couple of centuries before Newton came up with his theory of gravity, Copernicus must have had difficulty explaining why people—and everything else—clung to Earth's surface rather than flying off into space as the Earth spun around. So he envisioned that the Earth, the Moon, and other large bodies each had its own center to which everything would fall.

The process of continually modifying a mental model of Earth's position among other bodies in space continues today in the work of modern astronomers. Astronomers cannot "experiment" with objects in space. However, they can observe the light from distant stars and galaxies to see how they move and how they relate to one another, and then build a mental model (often translated into a more precise mathematical model) that is consistent with what they see. It is not unusual for a new observation—or a new way of thinking about old observations—to cause astronomers to change their mental models of stars, galaxies, moons, planets, or even the Earth itself.

Your Students' Mental Models

One of the greatest resources you can tap into as a teacher is the different mental models held by your students. By engaging them in discussing their ideas related to these probes, students will have an opportunity to learn about other mental models besides their own. That realization—that there may be a different way to think about a situation—is a tremendously powerful way to get your students to question their own thinking.

A set of instructional materials that used a similar approach but predated the *Uncovering Student Ideas in Science* series was *Earth, Moon, and Stars* in the GEMS (Great Explorations in Math and Science) series (Sneider 1986). Students were given "thought experiments," using diagrams and questions similar to some of the probes in this book. The focus of the GEMS unit was the Earth's spherical shape and a simple idea of gravity. The students started by writing their answers to the probes, and then they met in small groups to discuss their ideas. Without the teacher having to explain the correct concepts, these discussions alone moved the students much closer to the scientific mental model of the Earth in space (Sneider and Pulos 1983; Sneider et al. 1986).

The *Earth, Moon, and Stars* curriculum was rigorously tested in a study by 17 teachers from 10 states who agreed to give pretests and posttests to students who participated in their classes (Sneider and Ohadi 1998). The study included 539 students in grades 4–9. Although all groups significantly increased their test scores, the youngest students, in grades 4 and

5, made the greatest gains. Students in control groups, who did not engage in these activities, did not increase their scores. In other words, students *can* change their mental models of the Earth's shape and gravity concept by discussing their ideas with each other. And that is what we want you to do with these probes—use them to stimulate discussion that can help your students change their mental models of the universe.

Learning From Probes and the New Science Education Standards

In July 2011, the National Research Council (NRC) released a report entitled *A Framework for K–12 Science Education: Practices, Crosscutting Concepts, and Core Ideas* (NRC 2011). This Framework is expected to provide a blueprint for new science standards that states will eventually adopt. It describes core ideas in the major fields of science, including astronomy, and also describes important capabilities that all students should acquire during their K–12 experience. Among these are the crosscutting concepts of systems and models, stability and change, and cause and effect, and the practices of developing and using models, constructing explanations, and arguing from evidence. All of these capabilities can be supported by engaging students in discussions around the probes. Eventually the Next Generation Science Standards (NGSS) will be developed, based on the Framework; the NGSS will also embody these concepts and abilities. Although the NGSS are not yet available, the probes and Teacher Notes in this book will very likely be consistent because we have taken into account the practices, core ideas, and crosscutting concepts described in the 2011 *Framework* report.

Stepping-Stones

The word *misconception* has been used a number of times in the preface and introduction. It's important to point out that from the perspectives of the students their ideas about the world are not "wrong." As Vosniadou and Brewer (1992, 1994) and many others have pointed out, the students' explanations for phenomena such as night and day or phases of the Moon are entirely consistent with their mental models of the Earth in space, among a host of other objects that they cannot directly see or handle. Some researchers have suggested using the term *preconception* or *alternative conception* rather than *misconception* to avoid labeling the students' ideas as "wrong."

We prefer to keep the term *misconception* and use it in a general way to indicate that it is an idea that is contrary to the modern scientific thinking that we want to promote. However, we emphasize that expressing these ideas—even though they may be wrong from a modern scientist's point of view—is a very important part of the learning process. In many cases your students will progress through a sequence of misconceptions before getting it right. As suggested by researcher Philip Sadler (1998), misconceptions might best be thought of as stepping-stones that are absolutely essential for helping our students gradually change their mental models, so they can share the modern scientific view of the universe.

With this book you now have a powerful set of tools to uncover the astronomy-related ideas your students bring to the classroom. Through students' writing and talk you will be able to follow their efforts to make meaning out of their everyday encounters with the natural world and concepts presented to them in school. Every student has his or her own unique approach to creating meaning in a learning situation. Whether or not a student's ideas change depends on the willingness of the student to accept new ways of looking at his or her natural world.

In other words, you cannot "fix" your students' misconceptions. However, by using these probes to formatively assess your students' current thinking, you will be in a much better position to create a path that moves students from where they are to where they need to be scientifically. We hope the probes in this book, along with the information in the Teacher Notes, will help you create a classroom environment that makes it safe and interesting to surface and discuss all students' ideas, so that they can reach for the next stepping-stone in their understanding of astronomy.

References

Guthrie, W. K. C. 1939. *Aristotle on the Heavens.* Loeb Classical Library. Cambridge, MA: Harvard University Press.

National Research Council (NRC). 2011. *A framework for K–12 science education: Practices, crosscutting concepts, and core ideas.* Washington, DC: National Academies Press.

Sadler, P. M. 1998. Psychometric models of student conceptions in science: Reconciling qualitative studies and distracter-driven assessment instruments. *Journal of Research in Science Teaching* 35 (3): 265–296.

Sneider, C. 1986. *Earth, Moon, and Stars: Teacher's Guide.* GEMS. Berkeley, CA: Lawrence Hall of Science, University of California, Berkeley.

Sneider, C., and M. Ohadi. 1998. Unraveling students' misconceptions about the Earth's shape and gravity. *Science Education* 82 (2): 265–284.

Sneider, C., and S. Pulos. 1983. Children's cosmographies: Understanding the Earth's shape and gravity. *Science Education* 67 (2): 205–221.

Sneider, C., S. Pulos, E. Freenor, J. Porter, and B. Templeton. 1986. Understanding the Earth's shape and gravity. *Learning* 14 (6): 43–47.

Vosniadou, S., and W. F. Brewer. 1992. Mental models of the Earth: A study of conceptual change in childhood. *Cognitive Psychology* 24: 535–585.

Vosniadou, S., and W. Brewer. 1994. Mental models of the day/night cycle. *Cognitive Science* 18: 123–183.

Section 1

The Nature of Planet Earth

Concept Matrix: The Nature of Planet Earth
Probes 1–6

PROBES	1. Is Earth Really "Round"?	2. Where Do People Live?	3. Falling Through the Earth	4. What Causes Night and Day?	5. The Two *Rs*	6. Where Did the Sun Go?
GRADE-LEVEL USE →	3–8	3–8	6–12	3–12	3–12	3–12
RELATED CONCEPTS ↓						
day-night cycle				X	X	X
Earth: gravity		X	X			
Earth: orbit					X	
Earth: seasons					X	
Earth: shape	X	X	X	X		X
Earth: spin				X	X	X

Teaching and Learning Considerations

A full understanding that the Earth we live on is a spherical planet in space is essential if your students are to make much sense of other things that they learn in science, such as phases of the Moon, the planets of the solar system, stars, and galaxies. An understanding of the nature of Planet Earth is also essential for understanding world geography, weather and climate, and such current science topics as space satellites and robotic probes to Venus and Mars or the discovery of planets around other stars.

However, the idea that the solid Earth beneath our feet is actually a sphere that is suspended in space presents a major challenge to children in elementary school. It is much easier for them to observe regularities in the sky, such as the apparent daily motion of the Sun and the Moon's changing shape, than it is to imagine that people live down beneath their feet, on the other side of the world. That is why these ideas need to be revisited at the upper elementary level—and for some students at the middle and high school level, too—to ensure that they fully understand the implications of the spherical Earth concept.

The first three probes focus on the Earth itself—on its shape, on where people live around the world, and on the idea that we are held to Earth's surface by a mysterious force called gravity. The last three probes in this section concern the ideas that were developed by Copernicus in the 16th century—that the Earth spins on its axis every 24 hours and revolves around the Sun once a year.

Related Curriculum Topic Study Guides*

Earth's Gravity

Earth, Moon, Sun System

Motion of Planets, Moons, and Stars

*These guides are in Keeley, P. 2005. *Science Curriculum Topic Study: Bridging the Gap Between Standards and Practice.* Thousand Oaks, CA: Corwin Press and Arlington, VA: NSTA Press. Each Curriculum Topic Study Guide provides a process to help the reader (1) identify adult content knowledge, (2) consider instructional implications, (3) identify concepts and specific ideas, (4) examine research on learning, (5) examine coherency and articulation, and (6) clarify state standards and district curriculum.

Related NSTA and Other Resources

NSTA Press Books

American Association for the Advancement of Science (AAAS). 2001. *Atlas of science literacy.* Vol. 1. (See "Gravity" map, pp. 42–43, and "Solar System" map, pp. 43–44.) Washington, DC: AAAS.

Ansberry, K., and E. Morgan. 2010. *Picture-perfect science lessons: Using children's books to guide inquiry, 3–6.* (See "Day and Night," pp. 263–275.) Arlington, VA: NSTA Press.

Gilbert, S. 2011. *Models-based science teaching.* Arlington, VA: NSTA Press.

Holt, G., and N. West. 2011. *Project Earth science: Astronomy.* 2nd ed. Arlington, VA: NSTA Press.

Keeley, P., F. Eberle, and L. Farrin. 2005. *Uncovering student ideas in science, vol. 1: 25 formative assessment probes.* (See "Talking About Gravity," pp. 97–102.) Arlington, VA: NSTA Press.

Keeley, P., F. Eberle, and J. Tugel. 2007. *Uncovering student ideas in science, vol. 2: 25 more formative assessment probes.* (See "Darkness at Night," pp. 171–176.) Arlington, VA: NSTA Press.

Keeley, P., and R. Harrington. 2010. *Uncovering student ideas in physical science, vol. 1: 45 new force and motion assessment probes.* Arlington, VA: NSTA Press.

Koba, S., with C. T. Mitchell. 2011. *Hard-to-teach science concepts: A framework to support learners, grades 3–5.* (See Chapter 5, "Earth's Shape and Gravity.") Arlington, VA: NSTA Press.

Robertson, W. 2002. *Force and motion: Stop faking it! Finally understanding science so you can teach it.* Arlington, VA: NSTA Press.

Sneider, C. 2003. Examining students' work. In *Everyday assessment in the science classroom,* eds. J. M. Atkin and J. E. Coffey, 27–40. Arlington, VA: NSTA Press.

NSTA Journal Articles

Lightman, A., and P. Sadler. 1988. The Earth is round? Who are you kidding? *Science and Children* 25 (5): 24–26.

Nelson, G. 2004. What is gravity? *Science and Children* 42 (1): 22–23.

Philips, W. C. 1991. Earth science misconceptions. *The Science Teacher* 58 (2): 21–23.

NSTA Learning Center Resources

NSTA SciGuides

http://learningcenter.nsta.org/products/sciguides.aspx

Earth and Sky (K–4 and 5–8)

Gravity and Orbits

NSTA SciPacks

http://learningcenter.nsta.org/products/scipacks.aspx

Earth, Sun, and Moon

NSTA Science Objects

http://learningcenter.nsta.org/products/science_objects.aspx

Earth, Sun, and Moon: General Characteristics of Earth

Earth, Sun, and Moon: Our Moving Earth

Solar System: The Earth in Space

Other Resources

Agan, L., and C. Sneider. 2003. Learning about the Earth's shape and gravity: A guide for teachers and curriculum developers. *Astronomy Education Review* 2 (2): 90. *http://aer.aas.org/resource/1/aerscz/v2/i2/p90_s1*

Fraknoi, A., ed. 2011. *Universe at your fingertips 2.0.* San Francisco: Astronomical Society of the Pacific. Available at *www.astrosociety.org/uayf/index.html*

Sneider, C. 1986. *Earth, Moon, and stars: Teacher's guide.* GEMS. Berkeley, CA: Lawrence Hall of Science, University of California, Berkeley.

Is Earth Really "Round"?

Five friends were talking about the shape of the Earth. They each agreed the Earth is round. However, they disagreed about what "round" really means. Here are their ideas about a "round" Earth:

Chuck: "I read somewhere that Columbus or Magellan or someone proved the Earth is round like a round island. He sailed all around the island, and came back to the same port."

Sara: "I know the Earth doesn't look round. That's just because we live in a flat area. Other people can see it's round because they live near mountains and hills."

Takesha: "'Round' means that the whole Earth is shaped like a ball. It just looks flat because we can only see a small part of the ball."

Arnold: "You're right that 'round' means 'round like a ball,' but it looks flat because we live on the flat part in the middle. The upper part of the ball is the sky, and the bottom part is the solid Earth, where people live."

Missy: "Everyone knows that the round Earth is a planet in the solar system, like Mars and Jupiter. People get mixed up because 'earth' is also another name for the ground."

Who do you think has the best explanation? ____________ Explain why you think it is the best explanation, and use a drawing to support your explanation.

____

____

Include a drawing on the back of this page.

Is Earth Really "Round"?

Teacher Notes

Purpose

The purpose of this assessment probe is to elicit students' ideas about the shape of the Earth. The probe is designed to find out how students reconcile the idea that the Earth is round, represented by the globe that they are taught about in school, with the flat Earth of their everyday experience.

Related Concepts

Earth: shape

Explanation

Takesha has the best answer: "'Round' means that the whole Earth is shaped like a ball. It just looks flat because we can only see a small part of the ball." The size of the Earth in relation to where a person stands on the surface accounts for why we cannot see the Earth's curvature. From any point on the ground, you can only see a very small portion of the Earth in all directions; therefore, the portion you do see appears to be flat (uncurved). The higher you go beyond the surface of the Earth, the more you begin to see the curvature. As astronauts travel in a rocket or the Space Shuttle, away from Earth, they see the curved surface. Astronauts on the Moon, far from Earth, viewed Earth as a sphere. Many of the satellite images we have of the Earth from space shows the Earth to be shaped like a sphere.

Administering the Probe

This probe is best used with upper elementary and middle school students, since that is the age range when most students will have developed the spatial reasoning skills that will enable them to fully understand Takesha's argument. However, students as early as first or second grade will likely find the probe engaging and thought-provoking. Note that this probe asks students to support their thinking with a drawing. Consider having students share their drawings with the class to support their ideas about a "round Earth." Listen carefully as they explain their drawings. Some students might

reveal their thinking that there are actually two Earths—the one we live on and the one that is represented by a globe.

Related Ideas in *Benchmarks for Science Literacy* (AAAS 2009)

3–5 The Earth

- Things on or near the Earth are pulled toward it by the Earth's gravity.
- ★ The Earth is approximately spherical in shape.

Related Ideas in *National Science Education Standards* (NRC 1996)

5–8 Earth in the Solar System

- Gravity alone holds us to the Earth's surface and explains the phenomena of the tides.

Related Research

- In a series of studies in the United States and Israel, Joseph Nussbaum and his colleagues (Nussbaum 1979; Nussbaum and Novak 1976; Nussbaum and Sharoni-Dagan 1983) found that elementary students had great difficulty understanding the idea that the apparently flat ground beneath their feet is part of a huge ball in space, with no absolute "down" direction, except toward the center of Earth. Furthermore, to make sense of science lessons in which they are told that the Earth is "round," many students create their own explanations for what a "round Earth" means. Some of those alternative conceptions are expressed by Takesha's friends in this probe.
- A review of research studies (Agan and Sneider 2003) showed that Nussbaum and Novak's findings have been replicated many times and that students in many different countries share some of the same misconceptions, as do many teachers. While older students tend to give more scientifically accurate explanations to reconcile the spherical Earth concept with the apparently flat view that they can see every day, a significant number of students express alternative views as late as middle school. Fortunately, various teaching approaches have been found to be successful. The review recommended that a full presentation of these ideas should be delayed until at least fourth grade, when students have the capacity to understand the complex spatial thinking needed to reconcile the spherical Earth concept with the apparently flat ground of their everyday experience.
- A number of researchers have conducted learning studies to teach the spherical Earth concept. One of these studies (Sneider and Ohadi 1998) involved 17 teachers in 10 states who used the GEMS (Great Explorations in Math and Science) unit Earth, Moon, and Stars (Sneider 1986). The study included 539 students in grades 4–9. Although all groups significantly increased their test scores, the youngest students, in grades 4 and 5, made the greatest gains. Students in control groups, who did not engage in these activities, did not increase their understanding of the Earth's shape and gravity.

★ Indicates a strong match between the ideas elicited by the probe and a national standard's learning goal.

Suggestions for Instruction and Assessment

- Provide an opportunity for students to share, discuss, and critique their ideas about the probe while refraining from giving them the scientific answer. At a point when students are ready for a scientific explanation, give your own reasons for it. For example, explain how astronauts have now flown all the way around the Earth and have seen it as a ball in space.
- This probe can be combined with Probe 2, "Where Do People Live?"
- With older students, this probe provides an opportunity for students to use evidence to support their ideas about a spherical Earth. Students can examine some of the historical evidence for a spherical Earth well before humans had images of the Earth from space, and they can engage in argumentation using this evidence. Ask students to describe what they think this early historical evidence was and to discuss the strengths and weaknesses of the evidence. The discussion should cover the following evidence:
    - People at sea can see the tops of the masts of tall ships before seeing the hull of the ship or the whole mountain.
    - The Earth casts a circular shadow on the Moon during a lunar eclipse (although one may argue a flat disc would do this as well).
    - It is possible to circumnavigate the world—leaving from and returning to the same spot.
    - As one travels north, the Sun is lower in the sky, but stars such as Polaris (the north star) are higher in the sky. Other bright stars, such as Canopus (visible in Egypt), disappear from the sky.
    - The local time for a lunar eclipse is many hours later in the east (e.g., India) than in the west (e.g., Europe) even though the eclipse is seen by all observers simultaneously.
- A representation that may help students envision why it is hard to see the Earth as a sphere from their perspective is to compare large and small spheres. For example, looking at a Ping-Pong ball or other small sphere close up one can clearly see that it is curved. However, looking at a very large sphere, like a beach ball close up, one sees that it is less curved. The larger a sphere gets, the "flatter" its surface appears. Have students imagine why the Earth appears flat based on their observations of different-size spheres.
- Have students imagine what an ant's perspective would be when crawling on a beach ball and then make the link to the "flatness" of a spherical Earth from a human's perspective (Lightman and Sadler 1988).
- Show children a large spherical fruit such as a cantaloupe. Have students describe the overall shape of the melon. Then cut out a small square (about a square inch) of the melon's rind and have the students describe the shape of that small piece of the melon's surface. The small piece of rind will look flat to them. Have them link looking at the part of the Earth on which they are standing (e.g., the small square of melon rind) to the whole Earth (the entire melon) to explain why a part of a sphere appears flat (Hayes et al. 2003).
- The zooms on this NASA website help students see how the ground they stand on is part of a spherical Earth: *http://svs.gsfc.nasa.gov/search/Series/GreatZooms.html.* Similar zooming can be done with Google Earth.

References

Agan, L., and C. Sneider. 2003. Learning about the Earth's shape and gravity: A guide for teachers and curriculum developers. *Astronomy Education Review* 2 (2): 90. *http://aer.aas.org/resource/1/aerscz/v2/i2/p90_s1*

American Association for the Advancement of Science (AAAS). 2009. Benchmarks for science literacy online. *www.project2061.org/publications/bsl/online*

Hayes, B, A. Goodhew, E. Heit, and J. Gillan. 2003. The role of diverse instruction in conceptual change. *Journal of Experimental Child Psychology* 86: 253–276.

Lightman, A., and P. Sadler. 1988. The Earth is round? Who are you kidding? *Science and Children* 25 (5): 24–26.

National Research Council (NRC). 1996. *National science education standards.* Washington, DC: National Academies Press.

Nussbaum, J. 1979. Children's conceptions of the Earth as a cosmic body: A cross age study. *Science Education* 63 (1): 83–93.

Nussbaum, J., and J. Novak. 1976. An assessment of children's concepts of the Earth utilizing structured interviews. *Science Education* 60 (4): 535–550.

Nussbaum, J., and N. Sharoni-Dagan. 1983. Changes in second grade children's preconceptions about the Earth as a cosmic body resulting from a short series of audio-tutorial lessons. *Science Education* 67 (1): 99–114.

Sneider, C. 1986. *Earth, Moon, and stars: Teacher's guide.* GEMS. Berkeley, CA: Lawrence Hall of Science, University of California, Berkeley.

Sneider, C., and M. Ohadi. 1998. Unraveling students' misconceptions about the Earth's shape and gravity. *Science Education* 82 (2): 265–284.

Where Do People Live?

Three friends were arguing about what they would see if they could look straight through the Earth using x-ray vision. Here is what they said.

Jimmy: "If I looked down I would see people on the other side of the Earth. In fact, I would see the bottoms of their shoes."

Farouk: "I agree with Jimmy that we would see people on the other side of the Earth. But I think we would see the tops of their heads."

Sadi: "I disagree with both of you. There is nothing down under us except for dirt and rocks."

Which friend do you agree with the most? _________________ Explain why you agree with that friend and not the others.

____

____

____

____

____

Where Do People Live?

Teacher Notes

Purpose

The purpose of this assessment probe is to elicit students' ideas about the shape of the Earth. The probe is designed to find out if students grasp the idea that the spherical Earth applies to the actual Earth beneath their feet.

Related Concepts

Earth: gravity, shape

Explanation

Jimmy has the best answer: "If I looked down I would see people on the other side of the Earth. In fact, I would see the bottoms of their shoes." Because the Earth is spherically shaped, people can stand anywhere on the solid part of the sphere. If the Earth were transparent and you were able to look to the opposite side of the Earth to see other people standing, you would see the soles of their shoes or bottoms of their feet. Earth's gravity also explains why we would see the bottoms of people's shoes if we were able to look straight through the Earth. Gravity is a center-directed force that pulls objects toward the center of the Earth and accounts for why we can stand anywhere on Earth's spherical surface without being upside down or falling off the Earth.

Administering the Probe

This probe is primarily designed for students in upper elementary and middle grades, because around fourth grade most students will have developed the spatial reasoning skills that will enable them to envision looking all the way through a spherical Earth. Tell students that this type of probe is a "thought experiment" since it is impossible to look straight through the Earth and x-ray glasses do not really exist. Keep in mind that getting the "right answer" is not as important as the reasoning skills students develop and use to explain their mental model of the Earth in this imaginary situation. Few adults have ever been asked this question, so it can also be used to spur discussion during teacher workshops on elementary astronomy.

Related Ideas in *Benchmarks for Science Literacy* (AAAS 2009)

3–5 The Earth

- Things on or near the Earth are pulled toward it by the Earth's gravity.
- ★ The Earth is approximately spherical in shape.

6–8 The Earth

- Everything on or anywhere near the Earth is pulled toward the Earth's center by gravitational force.

Related Ideas in *National Science Education Standards* (NRC 1996)

5–8 Earth in the Solar System

- Gravity alone holds us to the Earth's surface and explains the phenomena of the tides.

Related Research

- In a replication of Nussbaum's original research on children's ideas about Earth's shape and gravity (Nussbaum 1979; Nussbaum and Novak 1976; Nussbaum and Sharoni-Dagan 1983), Sneider and Pulos (1983) and Sneider et al. (1986) found that even many middle school children had difficulty with a question similar to the one in this probe. One ninth grader, for example, insisted that you would only see "dirt and rocks" if you looked down through the Earth, and maybe "the devil."
- In a related question, about which way a person would look to see people in China if they could look through the Earth, most children and many adults pointed toward the east, parallel to the Earth. Even when challenged to say which way they would look in a straight line, not which way they would fly to get there, most insisted that you would look due east, suggesting a flat Earth model.

Suggestions for Instruction and Assessment

- This probe can be combined with Probe 1, "Is Earth Really 'Round'"?
- Middle school topics such as the solar system assume that students have a good grasp of the spherical Earth concept. However, researchers have found that as many as one in four middle school students are still confused about the Earth's shape and gravity, so you may want to use this probe to see if additional instruction about the Earth is necessary before students study the other planets.
- Clear plastic beach balls (or inflatable Earth globes if available) can be valuable tools in helping children envision what it is like to look downward, through the Earth. It is best if you have enough to distribute one globe to each small group of students so they can use it to share their ideas. If you have small, plastic figures of people, ask students to position them standing on the globe and observe them from the other side of the Earth globe.
- After the students have had a chance to discuss their ideas about this probe and to use a globe to further explore their ideas, lead a class discussion. You may find that more students now hold the scientific idea that you would see the bottoms of people's feet if you could look right through the Earth. Begin by asking the students what

★ Indicates a strong match between the ideas elicited by the probe and a national standard's learning goal.

they believe now and why, so that when you reveal the scientist's model, you will be able to address the students' current thinking. For example, if your students cannot envision someone living "down under" in Australia, you can turn an Earth globe around so that Australia is on top, and place a stick figure or plastic figurine of a person on Australia to demonstrate how Australians see the world.

- Point out to students that the convention of having north at the top of a map and east at the right on most modern maps was established by the astronomer Ptolemy nearly 2,000 years ago, and this convention was almost universally adopted by cartographers (Sprague 2001–2011). Discuss the notion of a "northern hemisphere bias" and the idea that "north" and "up" are not synonymous and that there is no absolute down. Have students do a Google search to see if they can find maps of the world that do not show the North Pole at the top.
- Students can explore antipodal points (points on the opposite side of the Earth) using this simulation: *http://astro.unl.edu/classaction/animations/coordsmotion/antipodesexplorer.html*

References

American Association for the Advancement of Science (AAAS). 2009. Benchmarks for science literacy online. *www.project2061.org/publications/bsl/online*

National Research Council (NRC). 1996. *National science education standards.* Washington, DC: National Academies Press.

Nussbaum, J. 1979. Children's conceptions of the Earth as a cosmic body: A cross age study. *Science Education* 63 (1): 83–93.

Nussbaum, J., and J. Novak. 1976. An assessment of children's concepts of the Earth utilizing structured interviews. *Science Education* 60 (4): 535–550.

Nussbaum, J., and N. Sharoni-Dagan. 1983. Changes in second grade children's preconceptions about the Earth as a cosmic body resulting from a short series of audio-tutorial lessons. *Science Education* 67 (1): 99–114.

Sneider, C., and S. Pulos. 1983. Children's cosmographies: Understanding the Earth's shape and gravity. *Science Education* 67 (2): 205–221.

Sneider, C., S. Pulos, E. Freenor, J. Porter, and B. Templeton. 1986. Understanding the Earth's shape and gravity. *Learning* 14 (6): 43–47.

Sprague, B. 2001–2011. Claudius Ptolemaeus (Ptolemy): Representation, understanding, and mathematical labeling of the spherical Earth. Center for Spatially Integrated Social Science, University of California, Santa Barbara. *www.csiss.org/classics/content/76*

Falling Through the Earth

A teacher asked her students to imagine that it was possible to drill a hole all the way through the Earth from the North Pole to the South Pole. The hole is lined with super-strong steel so that it does not collapse or melt. There is air inside the hole. She asked the students to discuss what would happen to a rock that is dropped into the hole. Here is what they said:

Alana: "It would fall into the hole and would just keep going until it hit something."

Nate: "It would just fall straight down and come out the other side."

Tess: "I bet it would come out the bottom of the Earth and just keep falling forever into space."

Tim: "It will go to the center of the hole and stop."

Jean: "It will pass through the center, slow down, and fall back toward the center again."

Frank: "It's probably just going to stick to the side somewhere." Whom do you agree with the most? ____________ Explain why you agree.

____

____

____

Falling Through the Earth

Teacher Notes

Purpose

The purpose of this assessment probe is to elicit students' ideas about Earth's gravity. The probe is designed to find out how students think gravity acts *inside* the Earth.

Related Concepts

Earth: gravity, shape

Explanation

Jean has the best answer to this thought experiment: "It will pass through the center, slow down, and fall back toward the center again." As soon as the rock is dropped, it starts accelerating. If there were no air in the hole it would keep accelerating until it passes Earth's center. Since the hole has air in it, the rock would reach a maximum speed fairly soon due to friction with the air, just like any falling object.

Many people get confused about what happens when the rock gets to the center of the Earth. That's usually because of a common misunderstanding of the term *center of gravity.* In fact, according to Newton's theory of gravity—accepted by virtually all scientists today—every particle in the Earth will attract every particle in the rock. So the rock has a net force toward Earth's center all the way down. (The term *center of gravity* is used to mean the point within an object around which mass is evenly distributed; so Earth's center of gravity is in the center of Earth. You can find the center of gravity of a small object by finding its balance point.)

What happens when the rock passes the center? The hole goes all the way through so there is nothing to stop the rock when it gets to the center. The rock is still falling. After passing

the Earth's center, the mass of the Earth is now pulling the rock back toward Earth's center, so it starts to slow down. As Jean says, it slows down, and then falls back toward the center. Eventually friction with the air in the hole will slow it down enough so it will stay suspended in the middle of the hole, balanced by forces pulling it in all directions.

Administering the Probe

This probe is best used with middle and high school students. It is best to administer this probe after students have had an opportunity to discuss and develop the scientific answers to Probes 1 and 2, because if they are still confused about Earth's shape, the direction of gravity inside the Earth will not make much sense to them. This question is challenging for many teachers, and even high school and college physics teachers sometimes struggle with the details, such as how far the rock would go past the center if there were no air to slow it down. Make sure students understand that this is a thought experiment and that realistically a hole cannot be drilled from one side of the Earth to the other.

Related Ideas in *Benchmarks for Science Literacy* (AAAS 2009)

3–5 The Earth

- Things on or near the Earth are pulled toward it by the Earth's gravity.
- The Earth is approximately spherical in shape.

6–8 The Earth

★ Everything on or anywhere near the Earth is pulled toward the Earth's center by gravitational force.

Related Ideas in *National Science Education Standards* (NRC 1996)

5–8 Earth in the Solar System

- Gravity is the force that keeps planets in orbit around the Sun and governs the rest of the motion in the solar system. Gravity alone holds us to the Earth's surface and explains the phenomena of the tides.

9–12 Motions and Forces

- Gravitation is a universal force that each mass exerts on any other mass.

Related Research

- Joseph Nussbaum, who was the first researcher to uncover students' difficulty in understanding Earth's shape and gravity, discovered that most children gradually come to understand the gravity concept by developing progressively more sophisticated "notions." The first is that things fall "down," which is consistent with a flat Earth model. When students have advanced to understand that Earth is a sphere and that people live all around the sphere without falling off, they must also develop a more sophisticated model of gravity in which things fall "down" to the surface of Earth. But they may be confused about what might happen inside the Earth. Eventually, they develop the mental model of gravity pulling toward Earth's center (Nussbaum 1979; Nussbaum and Novak 1976; Nussbaum and Sharoni-Dagan 1983).
- Sneider and Pulos (1983) proposed separating children's ideas about the Earth's shape from their understanding of gravity.

★ Indicates a strong match between the ideas elicited by the probe and a national standard's learning goal.

For example, some children whose notion of gravity was that everything falls "down" in space conceived of Earth as a flat pancake, whereas other children who held the same notion of gravity conceived of Earth as a ball in space and believed that people live just on top of the ball, because people living on the bottom of the ball would fall off.

- A later study (Bar et al. 1994) showed that by sixth grade most students are able to think of a spherical Earth with people living all over its surface, but they still have an immature concept of gravity, the most common being that gravity is caused by air pressure "holding us down" on the Earth (Bar, Sneider, and Martimbeau 1997).

Suggestions for Instruction and Assessment

- It is recommended that students be formatively assessed on their ideas related to Probes 1 and 2 before using this probe.
- Although some students in upper elementary school may be able to grasp Newton's explanation of what will happen to the rock, this is primarily a middle school topic, since students will need additional conceptual resources to envision every particle in the rock being attracted to every particle in the Earth.
- The spherical Earth and gravity concepts are intimately linked. If students fail to understand that gravity pulls everything on or near Earth toward the center, they may have difficulty envisioning people living all around the world.
- This probe would likely provide an excellent focus for discussion in a middle school classroom. However, as with all probes, do not give students the answer right away. Students need time to discuss and work through their own ideas first. When you do get to the point where you need to provide students with a scientific explanation, you may want to put it into a historical context. The famous philosopher Aristotle would have said that the rock would stop in the middle. As long ago as 350 BC, Aristotle taught that the Earth is shaped like a sphere. However, he thought that the center of Earth was the center of the universe and that everything would fall to the center of the universe because that was its "natural resting place." In 1543 AD Copernicus published a book saying that Earth is not the center of the universe and that it travels around the Sun. However, he, too, thought there was something special about the center of each planet that caused things to go to the center; so he would also have said the rock would stop in the middle. It wasn't until Sir Isaac Newton published his theory of gravity in 1687 that people understood the rock would pass the center and fall back and forth in the imaginary hole.
- The GEMS guide *Earth, Moon, and Stars* (Sneider 1986) includes an activity related to this probe, along with supporting teacher background material for teaching the concept of gravity as a center-directed force.
- This probe can be used as a summative assessment task for high school physics students after they have studied gravity. Even teachers and college students enjoy the discussion about this probe.

References

American Association for the Advancement of Science (AAAS). 2009. Benchmarks for science literacy online. *www.project2061.org/publications/bsl/online*

Bar, V., C. Sneider, and N. Martimbeau. 1997. What research says: Is there gravity in space? *Science and Children* 34 (7): 38.

Bar, V., B. Zinn, R. Goldmuntz, and C. Sneider. 1994. Children's concepts about weight and free fall. *Science Education* 78 (2): 149.

Copernicus, N. 1543. *De revolutionibus orbium coelestium* (*On the Revolutions of the Heavenly Spheres.* Nuremberg, Germany: Johannes Petreius.

National Research Council (NRC). 1996. *National science education standards.* Washington, DC: National Academies Press.

Newton, I. 1687. *Philosophiæ Naturalis Principia Mathematica (The Principia).* London: S. Pepys Press.

Nussbaum, J. 1979. Children's conceptions of the Earth as a cosmic body: A cross age study. *Science Education* 63 (1): 83–93.

Nussbaum, J., and J. Novak. 1976. An assessment of children's concepts of the Earth utilizing structured interviews. *Science Education* 60 (4): 535–550.

Nussbaum, J., and N. Sharoni-Dagan. 1983. Changes in second grade children's preconceptions about the Earth as a cosmic body resulting from a short series of audio-tutorial lessons. *Science Education* 67 (1): 99–114.

Sneider, C. 1986. *Earth, Moon, and Stars: Teacher's Guide.* GEMS. Berkeley, CA: Lawrence Hall of Science, University of California, Berkeley.

Sneider, C., and S. Pulos. 1983. Children's cosmographies: Understanding the Earth's shape and gravity. *Science Education* 67 (2): 205–221.

What Causes Night and Day?

Five friends were talking about what causes night and day. They each had different ideas. This is what they said:

Twyla: "We have daytime because that's when the Sun comes up. When we have nighttime it is because the Moon comes out."

Marcel: "I think it is daytime when the Sun comes up. It is nighttime when the Sun goes down beneath the Earth."

Tishon: "The Sun goes around the Earth once a day. So when it comes on our side of the Earth we have daytime."

Ashok: "Earth spins around once a day and that is why we have daytime and nighttime."

Joe: "It must be due to something else."

Which friend do you agree with the most? __________ Explain why you agree with that friend and not the others.

____

____

____

What Causes Night and Day?

Teacher Notes

Purpose

The purpose of this assessment probe is to elicit students' ideas about the day-night cycle. The probe is designed to find out if students can relate the shape of the Earth and its spin to why we have day and night.

Related Concepts

Day-night cycle
Earth: shape, spin

Explanation

Ashok has the best idea: "Earth spins around once a day and that is why we have daytime and nighttime." The Earth makes one complete turn on its axis each day. It is this spin of the entire Earth that makes it appear that the Sun rises in the morning and sets in the afternoon or evening. In fact the Sun does not actually rise and set (nor do the Moon or stars). All of the objects we see in the sky only appear to move as they do because the Earth on which we are standing is slowly spinning in space.

Administering the Probe

This probe can be used with elementary, middle, and even high school students. Night and day are typically introduced to children at a very young age, but it is not likely that they will fully understand and remember the explanation for night and day until at least upper elementary, and for most, middle school age. It is also a good idea to give this probe to high school students before a unit on astronomy, to ensure they have a solid understand of Earth's motions before studying the more complex motions of the other bodies of the solar system and beyond.

Related Ideas in *Benchmarks for Science Literacy* (AAAS 2009)

K–2 The Universe

- The Sun can be seen only in the daytime, but the Moon can be seen sometimes at

night and sometimes during the day. The Sun, Moon, and stars all appear to move slowly across the sky.

3–5 The Earth

★ The rotation of the Earth on its axis every 24 hours produces the night-and-day cycle. To people on Earth, this turning of the planet makes it seem as though the Sun, Moon, planets, and stars are orbiting the Earth once a day.

Related Ideas in *National Science Education Standards* (NRC 1996)

K–4 Objects in the Sky

- The Sun, Moon, stars, clouds, birds, and airplanes all have properties, locations, and movements that can be observed and described.

K–4 Changes in Earth and Sky

★ Objects in the sky have patterns of movement.

Related Research

- Because the explanation for the daily cycle of light and dark has traditionally been taught at the early elementary grades, some researchers have attempted to teach the concept as early as preschool (ages 5 and 6). However, they have met with little success (Valanides, Gritsi, and Kampeza 2000).
- The two most common misconceptions are that the day-night cycle is caused by the Earth going around the Sun once a day and that it is caused by the Sun going around the Earth once a day (Danaia and McKinnon 2007).
- An insightful series of studies on children's conception of Earth as a sphere, and their subsequent explanation for day and night, involved interviews of children in first, third, and fifth grades (Brewer 2008; Vosniadou and Brewer 1993, 1994). All of the children had learned about the scientific explanation for day and night in their textbooks. The interviews revealed that the younger children explained day and night by extending their everyday experience with light. If the Sun is no longer visible, it must be hiding behind something, such as behind clouds, mountains, or the Moon. Older children provided explanations that were closer to the scientific model, but also somewhat muddled, such as that the Sun and Moon go around the Earth every day. Most of the fifth graders explained day and night as due to Earth's rotation but still had misconceptions, such as that the Sun and Moon are stationary on either side, with Earth turning in the middle.
- Mant and Summers (1993) interviewed primary school teachers in England. Although most could explain the day-night cycle in scientific terms, few could relate their explanations to observations of how the Sun appears in the sky. Some appeared to work backward from their explanation to describe what must be happening in the sky. That suggests it is important to have students first observe how the Sun changes its position during the daytime, before explaining why that happens from the viewpoint of a spinning Earth.

★ Indicates a strong match between the ideas elicited by the probe and a national standard's learning goal.

Suggestions for Instruction and Assessment

- When discussing the probe, encourage students to add to or extend the explanation of the friend they agree with most. Additionally, you might ask students to critique the ideas of the other friends.
- This probe can be combined with "Darkness at Night" in *Uncovering Student Ideas in Science, Vol. 2: 25 More Formative Assessment Probes* (Keeley, Eberle, and Tugel 2007).
- It may not make a lot of sense to teach students that day and night occur because the Earth spins on its axis once every 24 hours until students fully understand that the apparently flat, solid Earth beneath their feet is actually a huge ball in space. Most students acquire the ability to visualize that idea around fourth grade. Therefore, first- or second-grade textbook illustrations of night and day are not likely to make much sense to most students.
- Although it is too early to teach kindergarteners or first graders the explanation for day and night and expect them to explain it clearly, they can learn that the Sun is out during the day but not at night, and that it is the Sun that determines when day starts and ends. Nighttime is simply the absence of sunlight. It is also important for students in kindergarten, first grade, or second grade to observe that the position of the Sun changes during the day, from one side of the sky to the other.
- Once upper elementary students understand that the Earth is a sphere, they can come to understand the concept at a deeper level when they apply that understanding to the day-night cycle.
- The day-night cycle can be demonstrated in a number of ways. Small groups of students can be provided with globes and stick figures or small dolls that can be placed on the globe with clay or tape. Darken the room as much as possible and turn on a bright light in the center of the room, so the students can see that just one half of their globe is lighted. They can then turn the globe slowly, noting how the "person" standing on Earth is first in daylight and then experiences night.
- You can follow the globe experience by having the students simulate the spinning Earth with their heads. Have the students slowly turn in place to see the "sunrise" as they just start to see the light, then watch the Sun go from one side of their field of view to the other side, and finally see "sunset" as the Sun disappears on the other side of their view.
- Revisit this idea at the high school level, asking students to write their explanation of the day-night cycle. Demonstrations like those mentioned above can be used if formative assessment reveals students have misconceptions.

References

American Association for the Advancement of Science (AAAS). 2009. Benchmarks for science literacy online. *www.project2061.org/publications/bsl/online*

Brewer, W. F. 2008. Naïve theories of observational astronomy: Review, analysis, and theoretical implications. In *International handbook of research on conceptual change,* ed. S. Vosniadou, 155–204. New York: Routledge.

Danaia, L., and D. H. McKinnon. 2007. Common alternative astronomical conceptions encountered in junior secondary science classes: Why is this so? *Astronomy Education Review* 6 (2): 32–53. *http://aer.aas.org/resource/1/aerscz/v6/i2/p32_s1*

Keeley, P., F. Eberle, and J. Tugel, 2007. *Uncovering student ideas in science, vol. 2: 25 more formative assessment probes.* Arlington, VA: NSTA Press.

Mant, J., and M. Summers. 1993. Some primary-school teachers' understanding of the Earth's place in the universe. *Research Papers in Education* 8 (1): 101–129.

National Research Council (NRC). 1996. *National science education standards.* Washington, DC: National Academies Press.

Valanides, N., F. Gritsi, and M. Kampeza. 2000. Changing pre-school children's conceptions of the day/night cycle. *International Journal of Early Years Education* 8 (1): 27–39.

Vosniadou, S., and W. Brewer. 1993. Constraints on knowledge acquisition: Evidence from children's models of the Earth and day/night cycle. In *Proceedings of the Fifteenth Annual Conference of the Cognitive Science Society,* 1052–1057. Mahwah, NJ: Lawrence Erlbaum Associates.

Vosniadou, S., and W. Brewer. 1994. Mental models of the day/night cycle. *Cognitive Science* 18: 123–183.

The Two *Rs*

Different words are used to describe the motion of objects in space. Circle the answer that best describes the meaning of the words *rotate* and *revolve.*

A *rotate* means spin; *revolve* means spin

B *rotate* means spin; *revolve* means orbit

C *rotate* means orbit; *revolve* means orbit

D *rotate* means orbit; *revolve* means spin

Explain how these words describe Earth's motion. You may use a diagram to support your explanation.

____

____

____

____

____

____

____

The Two *Rs*

Teacher Notes

Purpose

The purpose of this word use probe is to elicit students' ideas about two commonly confused words in astronomy—*rotate* and *revolve.* The probe is designed to find out if students can conceptually distinguish between the two terms and how they use the terms to describe Earth's motions.

Related Concepts

Day-night cycle
Earth: orbit, seasons, spin

Explanation

The best answer is B: *rotate* means spin; *revolve* means orbit. More specifically, *rotate* means to spin around a central axis and *revolve* means to go in a circle, or orbit, around a central location. Rotation is used to describe the day-night cycle. Revolution is used to describe Earth's annual path around the Sun.

Administering the Probe

This probe is best used after students have encountered the words *rotate* or *rotation* and *revolve* or *revolution.* This probe can be given during instruction on Earth's motions in space to formatively assess students' understanding of the terminology used in the unit. It can also be given months after the ideas have been taught to determine how well students retain their understanding of the terminology and the difference between the two motions. However, the intent of this probe is not to emphasize vocabulary over conceptual understanding. Because these two words often get in the way of conceptual understanding, the probe is used to determine if students understand these words related to the motions they describe. However, be aware that students can give correct definitions without conceptual understanding. If students can describe Earth's motions accurately but misuse the words, this is less problematic than if students know the words but cannot accurately describe

the motions. The concept is more important than the terminology. Data may reveal that the concept needs to be taught (or re-taught) before students are asked to associate the word with the concept. A Frayer Model can also be used as a formative assessment classroom technique (FACT) to determine whether students conceptually understand the meaning of these words (Keeley 2008).

Related Ideas in *Benchmarks for Science Literacy* (AAAS 2009)

3–5 The Universe

★ The Earth is one of several planets that orbit the Sun, and the Moon orbits around the Earth.

3–5 The Earth

★ The rotation of the Earth on its axis every 24 hours produces the night-and-day cycle. To people on Earth, this turning of the planet makes it seem as though the Sun, Moon, planets, and stars are orbiting the Earth once a day.

6–8 The Universe

- Nine planets of very different size, composition, and surface features move around the Sun in nearly circular orbits. *[Note: This benchmark was written before Pluto was reclassified.]*

Related Ideas in *National Science Education Standards* (NRC 1996)

5–8 Earth in the Solar System

- Most objects in the solar system are in regular and predictable motion. Those motions explain such phenomena as the day, the year, phases of the Moon, and eclipses.

Related Research

- Although there was no formal research found specifically related to students' meaning for these two words, in the authors' experiences there have been numerous instances of students' (and teachers') confusion between these two terms. Teachers are encouraged to conduct their own classroom research related to these terms and share their findings on the Uncovering Student Ideas website: *www.uncoveringstudentideas.org.*

Suggestions for Instruction and Assessment

- Dozens of websites include comments on the difficulty that students experience when trying to distinguish between these two terms and then remember the distinction. Even scientists sometimes use the words interchangeably, and some dictionary websites give these two terms as syn-

★ Indicates a strong match between the ideas elicited by the probe and a national standard's learning goal.

onyms for each other. So it is not surprising that students have difficulty learning both the concepts of relative motion and the terms intended to distinguish one from the other.

- The similarities in the sounds of the words as well as the concepts can cause confusion. Yet both "R" terms are typically used in textbooks to describe the relative motions of the Earth, Sun, and Moon. For example: "Earth *rotates* on its axis every 24 hours and *revolves* around the Sun in 365¼ days."
- One approach preferred by some teachers and textbooks is simply to use the terms *spin* and *orbit* rather than *rotate* and *revolve,* so that a sentence about Earth's motions would read: "Earth spins on its axis once in 24 hours and orbits the Sun in 365¼ days." The writers of *Benchmarks for Science Literacy (*AAAS 2009) appear to favor this approach. The Benchmarks use the term *rotation* with respect to Earth's daily motion, but they only use the term *revolution* when referring to the Copernican Revolution. Although that approach helps students understand and distinguish between the two types of motion, it does not help them distinguish the two terms if they should run across them later.
- Several authors emphasize the value of movement to help students understand the two different types of motion at a conceptual and kinesthetic level. For example, Susan Griss, in her book *Minds in Motion: A Kinesthetic Approach to Teaching Elementary Curriculum*, suggests that teachers use the movement to support the teaching of elementary school subjects. For example, a teacher could have students stand up and "spin" to help them understand the difference between rotation and revolution (Griss 1998).
- The motions of the Earth, Sun, and Moon are typically taught at the elementary school level. However, research (see Teacher Notes in Probes 1–4) strongly indicates that until at least fourth grade few students have developed the spatial abilities to visualize a spherical Earth in space undergoing different kinds of motions that result in phenomena they see from an apparently flat ground. So even if students were taught the concepts of revolution and rotation at earlier grades, before starting a middle school unit on the solar system it would be a good idea to use this probe to see if your students understand the different meanings of the two terms. This applies at the high school level as well. By the time students reach high school, nearly all should understand that Earth spins on its axis once a day and orbits the Sun once a year—but they may not remember which of the words *revolution* and *rotation* go with which concept.
- The Capital Region Science Education Partnership (2002) lists the following situations that can be included in a formative assessment as a way to see if your students can distinguish between revolution and rotation:
    - Children walking in a circle around the farmer while playing "The Farmer in the Dell" (revolution)
    - A basketball player spinning a basketball on a finger (rotation)
    - The planet Mars making one complete turn around the Sun (revolution)
    - The Moon making one complete turn around the Earth (revolution)
    - A wheel spinning on a car (rotation)
    - A chicken turning on a rotisserie (rotation)
    - A CD playing music (rotation)
    - A train running around the base of a Christmas tree (revolution)

  - A child riding on a Ferris wheel or merry-go-round (revolution)
  - The movement of the Ferris wheel or merry-go-round itself (rotation)
- Here is a mnemonic way for students to remember that rotation refers to Earth's daily spin: The middle syllable of the word *rotation* (row-tay-shun) rhymes with "day."
- The difference between the two motions can be described using a basketball and a carousel (merry-go-round). Imagine you are standing on a carousel spinning a basketball (Earth) on your fingertip. The carousel is slowly turning around a music box (Sun) in the center. As you ride the carousel, the basketball (Earth) slowly makes a complete path (revolution) around the music box (Sun) while it is spinning on your finger (rotation).

References

American Association for the Advancement of Science (AAAS). 2009. Benchmarks for science literacy online. *www.project2061.org/publications/bsl/online*

Capital Region Science Education Partnership. 2002. Another perplexing pair … rotation and revolution. *www.crsep.org/PerplexingPairs/AnotherPerplexingPair.RotationandRevolution121102.pdf*

Griss, S. 1998. *Minds in motion: A kinesthetic approach to teaching elementary curriculum.* Portsmouth, NH: Heinemann.

Keeley, P. 2008. *Science formative assessment: 75 practical strategies for linking assessment, instruction, and learning.* Thousand Oaks, CA: Corwin Press and Arlington, VA: NSTA Press.

National Research Council (NRC). 1996. *National science education standards.* Washington, DC: National Academies Press.

Where Did the Sun Go?

Six friends were looking up at a dark, night sky filled with stars. They wondered where the Sun was. This is what they said:

Naomi: "I think dark clouds in the sky hide the Sun at night."

Travis: "I think the Sun is beneath the Earth during nighttime."

Cooper: "I think the Sun goes way up with the stars during the night."

Leila: "I think the Sun is on the other side of the Earth during the night."

Mayumi: "I think the Sun is on the other side of the Moon during the night."

Suzanne: "I think the Sun just stops shining at night."

Which friend do you agree with the most? ___________ Explain why you agree with that friend.

____

____

____

____

____

Where Did the Sun Go?

Teacher Notes

Purpose

The purpose of this assessment probe is to elicit students' ideas about the position of the Sun in relation to the Earth. The probe is designed to find out how students envision the Earth as a sphere in space and its relationship to the Sun by asking students where the Sun is at night.

Related Concepts

Day-night cycle
Earth: shape, spin

Explanation

Leila has the best answer: "I think the Sun is on the other side of the Earth during the night." Since the Earth spins on its axis once a day, sometimes we are on the side of the Earth facing the Sun, and sometimes we are on the side facing away from the Sun.

Some students may be attracted to Travis's idea that the Sun is *beneath* the Earth during nighttime. Although Travis's idea is closer to the scientific idea than some of the others, he is not thinking of the Earth as a sphere in space, but rather as an object with an absolute down direction—so that some celestial objects are above the Earth and others are beneath it. Some students might express a flat Earth viewpoint by talking about the Sun being "below the horizon" or "over the horizon" in their explanation.

Administering the Probe

This probe is most appropriate for students in the upper elementary grades when students are able to fully grasp the idea that the Earth they live on is a sphere and that gravity pulls things toward Earth's center; that is, there is no "absolute down" direction in space. However, middle school students and even high school students could benefit from a discussion if some of the students initially choose Travis's idea. For younger students or English-language learner (ELL) students, consider cutting the number of choices to four.

Related Ideas in *Benchmarks for Science Literacy* (AAAS 2009)

K–2 The Universe

- The Sun can be seen only in the daytime, but the Moon can be seen sometimes at night and sometimes during the day.

3–5 The Earth

★ The Earth is approximately spherical in shape.

★ The rotation of the earth on its axis every 24 hours produces the night-and-day cycle. To people on earth, this turning of the planet makes it seem as though the sun, moon, planets, and stars are orbiting the earth once a day.

3–5 Constancy and Change

- Some things in nature have a repeating pattern, such as the day-night cycle, the phases of the Moon, and seasons.

Related Ideas in *National Science Education Standards* (NRC 1996)

K–4 Objects in the Sky

- The Sun, Moon, stars, clouds, birds, and airplanes all have properties, locations, and movements that can be observed and described.

Related Research

- Klein (1982) conducted one of the first studies of whether or not students understood lessons about the Earth's spherical shape, gravity, and the causes of day and night, which were then part of the regular first- and second-grade curriculum in Minnesota. Interviews with 24 second graders revealed that very few of the children understood the concepts, and several expressed ideas such as "the Sun hides at night."
- Sharp and Sharp (2007) tested an experimental astronomy unit in England with students in the upper elementary grades (ages 9–10). Based on interviews with 31 children before and after the course, the researchers found that the number of students who could give a scientific explanation for night and day increased from 35% to 75%.
- Plummer (2008) interviewed a total of 60 children in grades 1, 3, and 8, and developed a learning progression for realistic expectations for developing students' ideas about the Sun's motions:
  - For grades K–1 the Sun rises and sets, and it is in the sky during the daytime but not at night.
  - For grades 2–3 the Sun rises, moves continuously through the sky, and sets on the opposite side of the sky.
  - For grades 4–5 the Sun is highest at noon but does not pass directly overhead. Also for grades 4–5 the length of the Sun's path and its highest point in the sky changes across the seasons.

Suggestions for Instruction and Assessment

- This probe can be combined with the probe "Where Do Stars Go?" from *Uncovering Student Ideas in Science, Vol. 3: Another 25 Formative Assessment Probes* (Keeley, Eberle, and Dorsey 2008).

★ Indicates a strong match between the ideas elicited by the probe and a national standard's learning goal.

- This probe can be used to distinguish three groups of students: those who have a naïve understanding of the Earth and Sun in space (such as "goes behind mountains" or "is hidden by clouds"); those who know that the Earth and Sun both exist as objects in space but still envision an absolute "up" and "down" direction in space; and those who have a more scientific mental model that the Sun is on the "other side" of the Earth when we cannot see it, rather than "underneath" the Earth. The probe does not, however, distinguish between students who understand that the Sun's apparent motions are caused by Earth's daily spin, and those who think that the Sun goes around the Earth.
- For students in the upper elementary grades (4–5) who have difficulty with this probe, you may wish to address Earth's spherical shape and gravity, using Probes 1–4 to assess their understanding and to spark discussions. Then revisit this probe to see if their ideas have changed.
- There is an important difference between understanding at an intellectual level that we live on a huge sphere in space, and actually being able to visualize sunset as an "Earth turn." So even middle school students who have a clear mental model of the Earth and Sun may initially choose Travis's idea that the Sun goes "beneath" the Earth at night. Consequently, this probe may spark an interesting discussion. If some students choose Travis's idea, that will open the door to other students who realize that on a spherical Earth the term "beneath" would mean that the Sun would be *inside* the Earth's surface, which doesn't make any sense in this context.

References

American Association for the Advancement of Science (AAAS). 2009. Benchmarks for science literacy online. *www.project2061.org/publications/bsl/online*

Keeley, P., F. Eberle, and C. Dorsey. 2008. *Uncovering student ideas in science, vol. 3: Another 25 formative assessment probes.* Arlington, VA: NSTA Press.

Klein, E. 1982. Children's concepts of the Earth and the Sun: A cross-cultural study. *Science Education* 66 (1): 95–107.

National Research Council (NRC). 1996. *National science education standards.* Washington, DC: National Academies Press.

Plummer, J. 2008. Students' development of astronomy concepts across time. *Astronomy Education Review* 7 (1): 139–148. *http://aer.aas.org/resource/1/aerscz/v7/i1/p139_s1*

Sharp, J. G., and J. C. Sharp. 2007. Beyond shape and gravity: Children's ideas about the Earth in space reconsidered. *Research Papers in Education* 22 (3): 363–401.

Section 2

The Sun-Earth System

Concept Matrix: The Sun-Earth System
Probes 7–15

PROBES	7. Sunrise to Sunset	8. No Shadow	9. What's Moving?	10. Pizza Sun	11. How Far Away Is the Sun?	12. Sunspots	13. Shorter Days in Winter	14. Changing Constellations	15. Why Is It Warmer in Summer?
GRADE-LEVEL USE →	1–8	3–8	3–8	3–12	3–12	6–12	6–12	6–12	6–12
RELATED CONCEPTS ↓									
apparent vs. actual size				X	X				
objects in the sky	X	X						X	
seasons: cause	X	X					X	X	X
seasons: constellations			X					X	X
seasons: length of day	X						X		X
solar system objects: orbits			X					X	X
solar system objects: spin	X	X	X			X	X	X	
Sun: altitude at noon	X	X							X
Sun: distance					X				X
Sun: location relative to Earth			X		X				
Sun: path in the sky	X	X					X		X
Sun: size				X					
Sun: surface features						X			

Teaching and Learning Considerations

The relationship between the Earth and the Sun governs our daily existence on this planet, from the time we wake and go to sleep, to the clothes we choose to wear in the morning. Being able to envision the Earth and the Sun in space is the key to understanding such fundamental ideas as why we experience night and day as well as seasons.

The first three probes in this section concerns the most fundamental observations and ideas about the Sun's relationship to the Earth; that the Sun follows a predictable path from sunrise to sunset, that the Sun is highest in the sky around (but rarely exactly) at noon, and that when the Sun is not in the sky it is somewhere below the horizon, on the other side of the ball-shaped Earth.

The next three probes concern the Sun itself—how big it is compared with the Earth, how far away it is from us, and that it has a surface with features (sunspots) that can be observed with very simple instruments. **[Safety note: Observations of the Sun should always be done safely, using a projection method or a reliable Sun filter.]**

The last three probes in this section concern why we experience seasons. This is perhaps the most challenging topic in astronomy that students encounter in middle school. It is also one of the most valuable parts of the curriculum since it requires students to bring together what they have learned about the Sun-Earth system with information on seasonal changes in climate, the changing length of the day, the different constellations that become visible at different times of the year, and the recognition that the Southern and Northern hemispheres experience summer and winter in different months. While these three probes do not assess all of the component ideas that your students will need to fully understand seasons, they do address three of the most important concepts—that different constellations are visible in the sky as Earth progresses in its annual orbit around the Sun; that the changing length of daylight corresponds to changes in the path of the Sun in the sky during the year; and that the changing path of the Sun is a result of the tilt of Earth's axis with respect to the plane of its orbit around the Sun.

Related Curriculum Topic Study Guides*

Earth, Moon, Sun System
Motion of Planets, Moons, and Stars
Seasons

*These guides are found in Keeley, P. 2005. *Science Curriculum Topic Study: Bridging the Gap Between Standards and Practice.* Thousand Oaks, CA: Corwin Press and Arlington, VA: NSTA Press. Each Curriculum Topic Study Guide provides a process to help the reader (1) identify adult content knowledge, (2) consider instructional implications, (3) identify concepts and specific ideas, (4) examine research on learning, (5) examine coherency and articulation, and (6) clarify state standards and district curriculum.

Related NSTA and Other Resources

NSTA Press Books

American Association for the Advancement of Science (AAAS). 2001. *Atlas of science literacy.* Vol. 1. (See "Solar System" map, pp. 43–44.) Washington, DC: AAAS.

American Association for the Advancement of Science (AAAS). 2010. *Atlas of science literacy.* Vol. 2. (See "Weather and Climate" map, pp. 20–21.) Washington, DC: AAAS.

Gilbert, S. 2011. *Models-based science teaching.* Arlington, VA: NSTA Press.

Holt, G., and N. West. 2011. *Project Earth science: Astronomy.* 2nd ed. Arlington, VA: NSTA Press.

Keeley, P., F. Eberle, and C. Dorsey. 2008. *Uncovering student ideas in science, vol. 3: Another 25 formative assessment probes.* (See "Summer Talk," pp. 177–184, and "Me and My Shadow," pp. 185–190.) Arlington, VA: NSTA Press.

Konicek-Moran, R. 2008. *Everyday science mysteries.* (See "Where Are the Acorns?" pp. 39–50.) Arlington, VA: NSTA Press.

Konicek-Moran, R. 2010. *Even more everyday science mysteries.* (See "Daylight Saving Time," pp. 63–70.) Arlington, VA: NSTA Press.

Konicek-Moran, R. 2011. *Yet more everyday science mysteries.* (See "Sunrise, Sunset," pp. 69–78.) Arlington, VA: NSTA Press.

Robertson, W. 2009. *Answers to science questions from the stop faking it guy.* (See "What Causes the Seasons?" pp. 93–96.) Arlington, VA: NSTA Press.

NSTA Journal Articles

Abisdris, G. 2003. Observing sunspots. *The Science Teacher* 70 (3): 68–69.

Ansberry, K., and E. Morgan. 2009. Teaching through tradebooks: Sunrise, sunset. *Science and Children* 46 (8): 14–16.

Barrows, L. 2007. Bringing light onto shadows. *Science and Children* 44 (9): 43–45.

Bogan, D., and D. Wood. 1997. Simulating Sun, Moon, and Earth patterns. *Science Scope* 21 (2): 46, 48.

Brunsell, E., and J. Marcks. 2007. Teaching for conceptual change in space science. *Science Scope* 30 (9): 20–23.

Governor, D., A. Cetin-Dindar, P. Doney, J. Harper, J. Jenkins, C. Ortiz-Blanco, and D. Tippins. 2009. Solar paths: An international and integrated look at the sun and seasons. *Science Scope* 32 (8): 22–31.

Hall, C. B., and V. Sampson. 2009. Inquiry, argumentation, and the phases of the Moon: Helping students learn important concepts and practices. *Science Scope* 32 (8): 16–21.

Hemingway, M. 2000. Our star, the Sun. *Science and Children* 38 (1): 48–51.

Lambert, J., and S. Sundburg. 2010. Science shorts: The reasons for seasons. *Science and Children* 47 (8): 68–70.

Lindgren, J. 2003. Science sampler: Why we have seasons and other common misconceptions. *Science Scope* 26 (4): 50–51.

Nock, G. 2007. Science sampler: How long is your day? *Science Scope* 30 (9): 46–49.

Robertson, W. 2007. Science 101: What causes the seasons? *Science and Children* 44 (5): 54–57.

Thomas, J. 2011. The reasons for the seasons. *The Science Teacher* 78 (4): 52–57.

NSTA Learning Center Resources

NSTA SciGuides

http://learningcenter.nsta.org/products/sciguides.aspx

Earth and Sky (K–4 and 5-8)
Gravity and Orbits
Solar System

NSTA SciPacks

http://learningcenter.nsta.org/products/scipacks.aspx

Earth, Sun, and Moon
Solar System

NSTA Science Objects

http://learningcenter.nsta.org/products/science_objects.aspx

Earth, Sun, and Moon: Earth's Seasons
Earth, Sun, and Moon: Our Moving Earth
Solar System: The Earth in Space

Other Resources

Fraknoi, A., ed. 2011. *Universe at your fingertips 2.0.* San Francisco: Astronomical Society of the Pacific. Available at *www.astrosociety.org/uayf/index.html*

Gould, A., C. Willard, and S. Pompea. 2000. *The real reasons for seasons: Sun-Earth connection.* GEMS. Berkeley, CA: Lawrence Hall of Science, University of California, Berkeley.

Shugrue, S.K. Astronomy with a stick: Daytime astronomy for elementary and middle school students. National Science Teachers Association. *www.nsta.org/publications/interactive/aws-din/aws.aspx*

Sneider, C., Bar, V., and Kavanagh, C. 2011. Learning about seasons: A guide for teachers and curriculum developers. *Astronomy Education Review* 10 (1). *http://aer.aas.org/resource/1/aerscz/v10/i1/p010103_s1*

Sunrise to Sunset

Two friends were talking about where the Sun is in the sky between sunrise and sunset. They each drew a picture to explain their ideas. Here is what they drew and said:

Avi: "I think the Sun rises on one side and sets on the other."

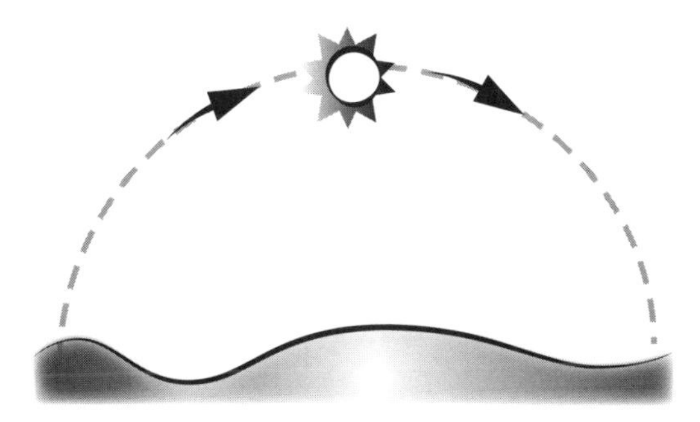

Jessica: "I think the Sun rises upward in the morning, then sets downward toward night. It looks like it goes up and down like this."

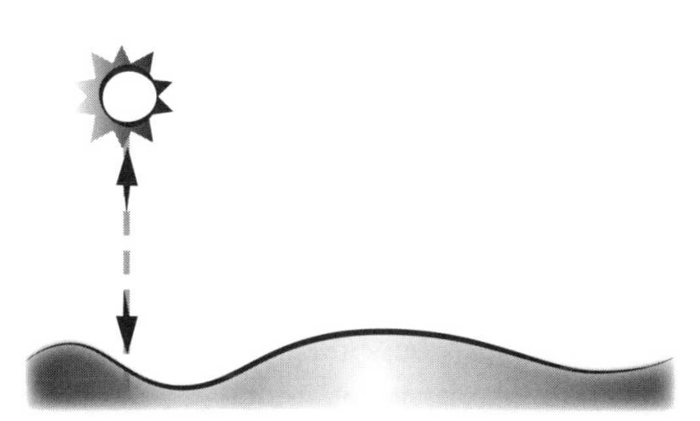

Whom do you agree with the most? ____________________ Explain why you agree.

____

____

____

____

____

____

Sunrise to Sunset

Teacher Notes

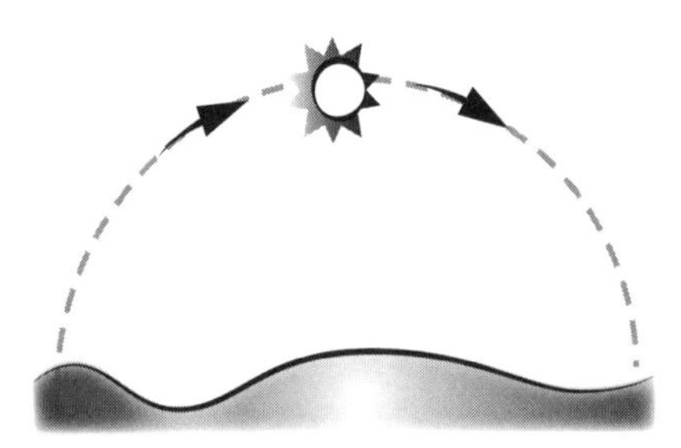

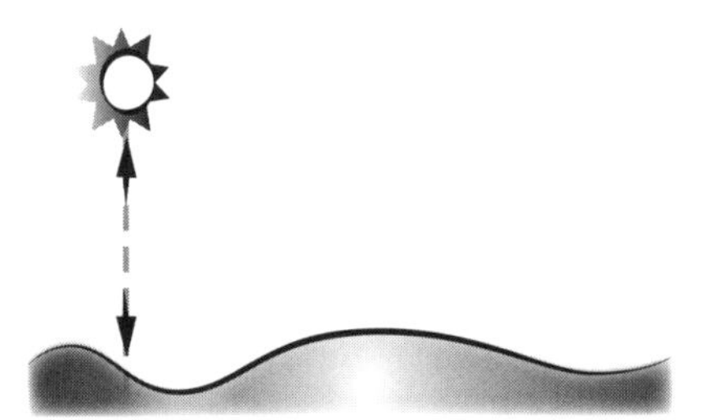

Purpose

The purpose of this assessment probe is to elicit students' ideas about the apparent movement of the Sun. The probe is designed to reveal where students think the Sun rises and sets, and how it moves during the day.

Related Concepts

Objects in the sky
Seasons: cause, length of day
Solar system objects: spin
Sun: altitude at noon, path in the sky

Explanation

Avi has the best answer. The Sun's apparent motion across the sky during the course of a day is arclike as shown in Avi's drawing. The Sun rises along the eastern horizon, appears to travel across the entire sky, and sets along the western horizon. Viewed from the Northern Hemisphere, facing south, it seems like the Sun is moving across the sky in a left-to-right direction. At around noon (which is the midpoint between sunrise and sunset) the Sun appears to be at its highest point in the sky. However, the Sun only *appears* to move in an arc across the sky. It is the Earth's rotation that is responsible for this visual effect.

Administering the Probe

This probe is primarily designed for students in the elementary and middle school grades. For the youngest children you may want to read the probe aloud and allow time for the children to ask questions. For middle school students you may want to add an additional task—to draw how the Sun appears to move through the sky during the day. **[Safety note: If you follow this probe with Sun observations throughout the day, make sure students do not look directly at the Sun.]**

Related Ideas in *Benchmarks for Science Literacy* (AAAS 2009)

K–2 The Universe

★ The Sun, Moon, and stars all appear to move slowly across the sky.

3–5 The Earth

- The rotation of the Earth on its axis every 24 hours produces the night-and-day cycle. To people on Earth, this turning of the planet makes it seem as though the Sun, Moon, planets, and stars are orbiting the Earth once a day.

Related Ideas in *National Science Education Standards* (NRC 1996)

K–4 Objects in the Sky

- The Sun, Moon, stars, clouds, birds, and airplanes all have properties, locations, and movements that can be observed and described.

K–4 Changes in Earth and Sky

★ Objects in the sky have patterns of movement.

Related Research

- Plummer (2008) interviewed 20 students in each of grades 1, 3, and 8. Although she found a general trend toward higher levels of understanding among the older students, students at each grade level held misconceptions about how the Sun appeared to move through the sky during the day and how the Sun's path across the sky changed with the seasons. Many of the children at all ages thought that the Sun was directly overhead at noon every day, even though the Sun was never overhead at noon at the latitude where the children lived. Furthermore, there was no significant difference between third-grade students' and eighth-grade students' understanding of the Sun's apparent motions.
- Plummer and Krajcik (2010) found that children as young as first grade knew that the Sun gets higher in the sky during the day and lower in the sky during the evening, although most were not able to accurately describe the Sun's path. Some of the students envisioned the Sun going up and then down on the same side of the sky. Other children thought the Sun stopped moving in the sky during the day. However, after a planetarium program about the Sun's path, 86% of first- and second-grade students were able to describe the Sun's path as rising on one side of the sky, following a continuous arc, and setting on the other side of the sky.
- Mant and Summers (1993) interviewed primary school teachers in England. Although most could explain the day-night cycle in scientific terms, few could relate their explanations to observations of how the Sun appears in the sky. Some appeared to work backward from their explanation to describe what must be happening in the sky. That suggests it is important to have students first observe how the Sun changes its position during the daytime, before explaining why that happens from the viewpoint of a spinning Earth.

Suggestions for Instruction and Assessment

- This probe can be combined with "Darkness at Night" in *Uncovering Student Ideas in Science, Vol. 2: 25 More Formative Assessment Probes* (Keeley, Eberle, and Tugel 2007).

★ Indicates a strong match between the ideas elicited by the probe and a national standard's learning goal.

- For students who believe that the Sun rises vertically above the horizon in the morning and then comes back down toward evening, be aware that these students might be bound by the way we use the words *sunup and sundown* or *sunrise and sunset,* which implies that the Sun literally goes up and down in the sky. It is particularly important that these students have an opportunity to observe the position of the Sun throughout the day. **[Safety note: Students should never look directly at the Sun.]** The best way to do that is to have them mark the position of a fixed object, like a flagpole. It is important to explicitly point out that the shadow marks the direction *opposite* the Sun. When the Sun is highest in the sky, the shadow will be the shortest.
- Although it is too early to teach kindergarteners or first graders the explanation for day and night and expect them to explain it clearly, they can learn that the Sun is out during the day but not at night and that it is the Sun that in fact determines when day starts and ends. Nighttime is simply the absence of sunlight. It is also important for students in kindergarten or first or second grade to observe that the position of the Sun changes during the day, moving in a smooth continuous arc from one side of the sky to the other.
- In upper elementary school, when students study the Earth as a ball in space, you can have the students simulate the spinning Earth with their heads by slowly turning to see the "sunrise" as they just start to see the light, watch the Sun go from one side of their field of view as they slowly turn, then see "sunset" as the Sun disappears on the other side of their view.
- For older students, the University of Oregon's Solar Radiation Monitoring Laboratory has a website where students can create Sun path charts from their location: *http://solardat.uoregon.edu/SunChartProgram.html*

References

American Association for the Advancement of Science (AAAS). 2009. Benchmarks for science literacy online. *www.project2061.org/publications/bsl/online*

Keeley, P., F. Eberle, and J. Tugel, 2007. *Uncovering student ideas in science, vol. 2: 25 more formative assessment probes.* Arlington, VA: NSTA Press.

Mant, J., and M. Summers. 1993. Some primary-school teachers' understanding of the Earth's place in the universe. *Research Papers in Education* 8 (1): 101–129.

National Research Council (NRC). 1996. *National science education standards.* Washington, DC: National Academies Press.

Plummer, J. 2008. Students' development of astronomy concepts across time. *Astronomy Education Review* 7 (1): 139–148. *http://aer.aas.org/resource/1/aerscz/v7/i1/p139_s1*

Plummer, J., and J. Krajcik. 2010. Building a learning progression for celestial motion: Elementary levels from an Earth-based perspective. *Journal of Research in Science Teaching* 47 (7): 768–787.

No Shadow

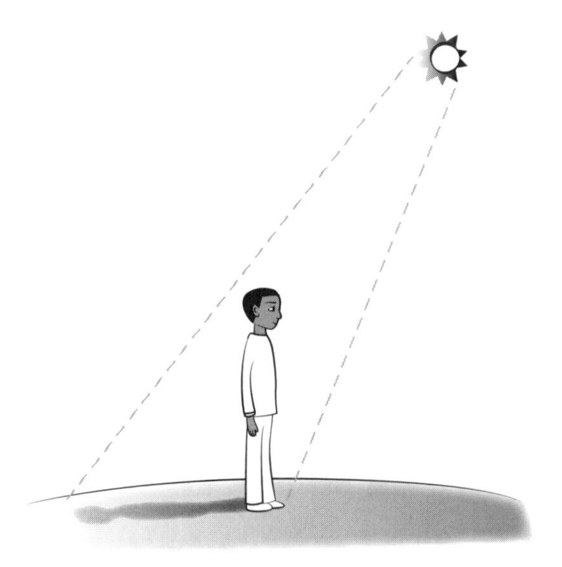

Four friends in New York City were wondering if there was ever a time during the day when they would have no shadow. They each had different ideas about daytime shadows. This is what they said:

Paige: "I will have no shadow when the Sun is highest in the sky."

Olive: "I will have no shadow at noon because the Sun will be directly overhead."

Kami: "It depends on the season. I will have no shadow at noon in the summer, but not in the winter."

Vince: "It depends on where you live. Only people who live near the equator will have no shadow on certain days."

Which friend do you agree with the most? ________________ Explain why you agree.

____

____

____

____

____

No Shadow

Teacher Notes

Purpose

The purpose of this assessment probe is to elicit students' ideas about the Sun's motion during the day. The probe is designed to reveal whether students know that the Sun is never directly overhead as viewed from the continental United States. (Hawaii is the only state where the Sun is sometimes overhead at noon.)

Related Concepts

Objects in the sky
Seasons: cause
Solar system objects: spin
Sun: altitude at noon, path in the sky

Explanation

Vince has the best answer: "It depends on where you live. Only people who live near the equator will have no shadow on certain days." In order to have no shadow, the Sun needs to be directly overhead. The Sun is never directly overhead except on certain occasions for locations between the Tropic of Cancer and the Tropic of Capricorn. The only state that falls between the tropics is Hawaii. So for the continental United States and Alaska there is always a shadow whenever the Sun is visible in the sky. Paige's response reflects a common misconception that the Sun is directly overhead when it is highest in the sky. Olive expresses a widely held misconception that the Sun is overhead at noon. Kami's statement that it depends on the season would be correct if she lived in the tropics, and even then the Sun is not overhead every day.

Administering the Probe

This probe is primarily designed for students in grades 3–8. For the younger children it is important to first determine if they understand how shadows form and that the length of their shadow indicates how high the Sun is in the sky. If some children are confused about the relationship between the length of their shadow and the height of the Sun, plan some preliminary activities for students to learn

about this relationship before administering the probe. For middle school students, you can extend the probe by asking the students to explain their answer. Students can also be encouraged to draw a picture to support their explanations.

Related Ideas in *Benchmarks for Science Literacy* (AAAS 2009)

K–2 The Universe

★ The Sun, Moon, and stars all appear to move slowly across the sky.

3–5 The Earth

- The rotation of the Earth on its axis every 24 hours produces the night-and-day cycle. To people on Earth, this turning of the planet makes it seem as though the Sun, Moon, planets, and stars are orbiting the Earth once a day.

Related Ideas in *National Science Education Standards* (NRC 1996)

K–4 Objects in the Sky

- The Sun, Moon, stars, clouds, birds, and airplanes all have properties, locations, and movements that can be observed and described.

K–4 Changes in Earth and Sky

- Objects in the sky have patterns of movement.

Related Research

- Plummer and Krajcik (2010) interviewed 60 students—20 each in grades 1, 3, and 8—about their ideas concerning how the Sun moves in the sky during the day. None of the first-grade students and only a few of the older students (20% in third grade and 10% in eighth grade) knew that the Sun does not pass directly overhead.
- A sample item from a set of astronomy diagnostic questions asks college students, "As seen from your location, when is the Sun directly overhead at NOON (so that no shadows are cast)? " Common incorrect responses included every day; on the day of the summer solstice; on the day of the winter solstice; and at both of the equinoxes (spring and fall). Many college students failed to select the correct response: never from the latitude of your location (Zeilik, Schau, and Matter 1998).
- Students seem to have more success in locating where an object's shadow will fall in relation to a light source if the object is a person. They have more difficulty anticipating where a shadow will fall if it is a nonhuman object, such as a tree (Driver et al. 1994).
- Plummer (2008) interviewed a total of 60 children in grades 1, 3, and 8, and developed a learning progression for realistic expectations for developing students' ideas about the Sun's motions:
  - For grades K–1 the Sun rises and sets, and it is in the sky during the daytime but not at night.
  - For grades 2–3 the Sun rises, moves continuously through the sky, and sets on the opposite side of the sky.
  - For grades 4–5 the Sun is highest at noon but does not pass directly overhead. Also for grades 4–5, the length

★ Indicates a strong match between the ideas elicited by the probe and a national standard's learning goal.

of the Sun's path and its highest point in the sky changes across the seasons.

- Mant and Summers (1993) interviewed primary school teachers in England. Although most could explain the day-night cycle in scientific terms, few could relate their explanations to observations of how the Sun appears in the sky. Some appeared to work backward from their explanation to describe what must be happening in the sky. That suggests it is important to have students first observe how the Sun changes its position during the daytime before explaining why that happens from the viewpoint of a spinning Earth.

Suggestions for Instruction and Assessment

- This probe can be combined with "Me and My Shadow" in *Uncovering Student Ideas in Science, Vol. 3: Another 25 Formative Assessment Probes* (Keeley, Eberle, and Dorsey 2008).
- This probe can be used to launch into an investigation in which students measure the length of a shadow from morning through afternoon, including observations at about 15-minute intervals before and after noon. This activity can help students learn why it is colder in winter if repeated monthly throughout the year. Record the observations on a large sheet of paper and post them on a wall so students can compare the observations; they will see that as winter approaches the Sun does not go as high in the sky around noon, so their shadow is longer. As the school year turns to spring, the Sun will be higher in the sky around noon so shadows will be shorter.
- Revisit this idea at the high school level, asking students to write an explanation of the day-night cycle. Demonstrations like those mentioned above can be used if formative assessment reveals students have misconceptions about the motion of the Sun during the day.
- NSTA's Astronomy With a Stick project has several activities and suggestions for helping students develop an understanding of the changing position of the Sun throughout the day: *www.nsta.org/publications/interactive/aws-din/aws.aspx.*

References

American Association for the Advancement of Science (AAAS). 2009. Benchmarks for science literacy online. *www.project2061.org/publications/bsl/online*

Driver, R., A. Squires, P. Rushworth, and V. Wood-Robinson. 1994. *Making sense of secondary science: Research into children's ideas.* London: Routledge.

Keeley, P., F. Eberle, and C. Dorsey. 2008. *Uncovering student ideas in science, vol. 3: Another 25 formative assessment probes.* Arlington, VA: NSTA Press.

Mant, J., and M. Summers. 1993. Some primary-school teachers' understanding of the Earth's place in the universe. *Research Papers in Education* 8 (1): 101–129.

National Research Council (NRC). 1996. *National science education standards.* Washington, DC: National Academies Press.

Plummer, J. 2008. Students' development of astronomy concepts across time. *Astronomy Education Review* 7 (1): 139–148. *http://aer.aas.org/resource/1/aerscz/v7/i1/p139_s1*

Plummer, J., and J. Krajcik. 2010. Building a learning progression for celestial motion: Elementary levels from an Earth-based perspective. *Journal of Research in Science Teaching* 47 (7): 768–787.

Zeilik, M., C. Schau, and N. Matter. 1998. Misconceptions and their change in university-level astronomy courses. *The Physics Teacher* 36: 104–107.

What's Moving?

Which statement best describes the movement of the Earth and Sun? Circle the answer that best matches your thinking.

A The Earth goes around the Sun once a day.

B The Sun goes around the Earth once a day.

C The Earth goes around the Sun once a year.

D The Sun goes around the Earth once a year. ______________________________

Explain your thinking. Describe the evidence that supports your answer.

____

____

____

____

____

____

What's Moving?

Teacher Notes

Purpose

The purpose of this assessment probe is to elicit students' ideas about Earth's orbit. The probe is designed to find out whether students recognize that the Earth orbits around the Sun.

Related Concepts

Seasons: constellations
Solar system objects: orbits, spin
Sun: location relative to Earth

Explanation

The best answer is C. The Earth goes around the Sun once a year. Evidence that supports a yearly cycle is that we see different constellations at different times of the year in a repeating annual cycle. An interesting historical note is that until the early 16th century the prevailing point of view was that the stars went around the Earth once a year. That was a reasonable supposition at the time, before people realized how huge and distant the stars were. Galileo's first telescope observations, showing that Jupiter had moons circling around it, was the first strong evidence that Copernicus was right—Earth and all of the other planets circled the Sun.

Administering the Probe

This probe is designed for elementary and middle school students but can be used at all grade levels. It can also be used with adults, who may know that the Earth orbits the Sun and may have been taught the evidence to support this idea, but may not have learned the details. Make sure students describe the evidence for their answer choice and not just state the fact; keep in mind that young students will provide evidence from their everyday experiences before they encounter the scientific explanation. You can ask students to extend their explanations by drawing a picture to support their ideas.

Related Ideas in *Benchmarks for Science Literacy* (AAAS 2009)

3–5 The Universe

- The patterns of stars in the sky stay the same, although they appear to move across the sky nightly, and different stars can be seen in different seasons.
- ★ The Earth is one of several planets that orbit the Sun, and the Moon orbits around the Earth.

6–8 The Universe

- Nine planets of very different size, composition, and surface features move around the Sun in nearly circular orbits. *[Note: This benchmark was written before Pluto was reclassified.]*

Related Ideas in *National Science Education Standards* (NRC 1996)

5–8 Earth in the Solar System

- ★ The Sun, an average star, is the central and largest body in the solar system.
- Most objects in the solar system are in regular and predictable motion. Those motions explain such phenomena as the day, the year, phases of the Moon, and eclipses.

Related Research

- As part of an evaluation of a high school astronomy course, Sadler (1998) tested 1,250 high school students on basic astronomical knowledge. About 66% of the students correctly answered the question "What causes night and day?" by choosing the response "The Earth spins on its axis." Most of the students who failed the item chose "The Earth moves around the Sun." What was most surprising was that as students learned more about astronomy, they were more likely to respond to this question by choosing a factually correct statement ("The Earth moves around the Sun") even though it was the wrong answer to the question (about the cause of night and day). In other words, the more students learned about motions of the Earth, Sun, and Moon, the more confused they became—at least until they were able to straighten it all out. Sadler concluded, "New curricula should discuss and treat alternative concepts not as errors, but as stepping stones to scientific understanding" (p. 290).
- Danaia and McKinnon (2007) tested 1,920 middle school students in grades 7, 8, and 9 in Australia on a number of different astronomy tasks. The percentage of correct responses to a question about the causes of the day-night cycle was 19.5% in grade 7, 31.5% in grade 8, and 39.9% in grade 9. The most common misconceptions were that "The Earth orbits the Sun daily" or "The Sun goes around the Earth every day." Although all students had learned about the motions of the Earth, the great majority of middle school students were confused about what went around what.

Suggestions for Instruction and Assessment

- Although students may be introduced to Earth's spin and its orbit around the Sun in elementary school, it is common for students to forget the details and later confuse what goes around what. The motions of the Earth, the Moon, and the planets

★ Indicates a strong match between the ideas elicited by the probe and a national standard's learning goal.

are not simple, especially when students are expected to envision these bodies moving in space in order to answer questions such as the causes of the day-night cycle, Moon phases, eclipses, and seasons. Even if students successfully learn these explanations, as the years go by they may forget the details and thus no longer recall for certain what goes around what.

- It may be better for students to learn about Earth's daily spin and its orbit around the Sun on subsequent days, so that they can compare these two different motions and recognize the phenomena that they explain.
- Commonly, students learn about Earth's orbit around the Sun as an explanation for the seasons. However, it is much easier for them to first understand how Earth's orbit around the Sun makes different constellations visible at different times of year than it is to explain the seasons. (See Probe 14, "Changing Constellations.")
- A way to engage students in thinking more deeply about what goes around what is to create diagrams of (1) the Sun orbiting the Earth, (2) the Earth spinning on its axis next to the Sun, (3) the Earth orbiting the Sun, and (4) the Earth spinning on its axis and orbiting the Sun. Ask students to work in teams and discuss which diagram best describes the motion of the Sun-Earth system and explain why.

References

American Association for the Advancement of Science (AAAS). 2009. Benchmarks for science literacy online. *www.project2061.org/publications/bsl/online*

Danaia, L., and D. H. McKinnon. 2007. Common alternative astronomical conceptions encountered in junior secondary science classes: Why is this so? *Astronomy Education Review* 6 (2): 32–53. *http://aer.aas.org/resource/1/aerscz/v6/i2/p32_s1*

National Research Council (NRC). 1996. *National science education standards.* Washington, DC: National Academies Press.

Sadler, P. M. 1998. Psychometric models of student conceptions in science: Reconciling qualitative studies and distracter-driven assessment instruments. *Journal of Research in Science Teaching* 35 (3): 265–296.

Pizza Sun

Imagine a pizza with slices of tomato, pepperoni, pieces of chopped green pepper, and grated cheese. If the pizza represents the Sun, circle what you think would best represent the Earth to show how big it is compared with the Sun.

A a round table that is about 10 times larger than the pizza, so that it would take 10 Suns to stretch across the Earth

B a slice of tomato that is about one-fifth the size of the pizza, so that it would take 5 Earths to stretch across the Sun

C a slice of pepperoni that is about one-tenth the size of the pizza, so that it would take 10 Earths to stretch across the Sun

D a piece of chopped green pepper that is about one-hundredth the size of the pizza, so that it would take 100 Earths to stretch across the Sun

E a speck of grated cheese that is about one-thousandth the size of the pizza, so that it would take 1,000 Earths to stretch across the Sun

Explain your thinking. How did you decide how big the Earth is compared with the Sun?

____

____

____

Pizza Sun

Teacher Notes

Purpose

The purpose of this assessment probe is to elicit students' ideas about relative size. The probe is designed to find out how large students think the Earth is compared with the Sun.

Related Concepts

Apparent vs. actual size
Sun: size

Explanation

The best answer is D: "A piece of chopped green pepper that is about one-hundredth the size of the pizza, so that it would take 100 Earths to stretch across the Sun." To be precise, Earth's diameter is 109 times smaller than the Sun's. Note that this question compares relative size when looking at the face of the Sun and the Earth (area). If the question were about volume—the number of Earths that could fit inside the Sun—the answer would be approximately 1.3 million.

Administering the Probe

This probe can be given to students at all grade levels. It can be used to find out students' preconceptions related to the size of the Sun and Earth, or it can be used following instruction about the solar system to see whether students developed a sense of scale size as a result of their instructional experiences. Make sure students know they are considering the diameter to relate the two sizes. If your students have difficulty understanding that the pizza and tomato slices represent cross sections of the Sun and Earth, you can illustrate this by slicing an apple and a grape in half and having them estimate how many grape diameters equal one apple diameter. If they have difficulty comparing the diameters of two discs, you can demonstrate with a paper plate and a coin. You may also want to create a diagram to show the size of each ingredient on the pizza.

If the fractional comparisons are too difficult for students, consider revising the wording as follows:

A a round table that is about 10 times bigger than the pizza
B a slice of tomato that is about 5 times smaller than the pizza
C a slice of pepperoni that is about 10 times smaller than the pizza
D a piece of chopped green pepper that is about 100 times smaller than the pizza
E a speck of grated cheese that is about 1,000 times smaller than the pizza

Related Ideas in *Benchmarks for Science Literacy* (AAAS 2009)

K–2 Scale

- Things in nature and things people make have very different sizes, weights, ages, and speeds.

3–5 The Universe

- The Earth is one of several planets that orbit the Sun, and the Moon orbits around the Earth.

Related Ideas in *National Science Education Standards* (NRC 1996)

K–4 Objects in the Sky

- The Sun, Moon, stars, clouds, birds, and airplanes all have properties, locations, and movements that can be observed and described.

Related Research

- Jones, Lynch, and Reesink (1987) interviewed 32 children from the third and sixth grades in Tasmania, asking them to pick out three-dimensional shapes and sizes to illustrate their understanding of the shapes and sizes of the Earth, the Sun, and the Moon. Only 25% of the children indicated that the Sun is larger than the Earth and that the Earth is larger than the Moon. Although there were no gender differences among the third graders, in sixth grade seven of the boys chose the correct order of sizes and only one girl chose the correct order.
- Sadler (1987) interviewed 25 ninth-grade students about their ideas concerning the Earth, the Sun, and the Moon. About half of the students had just taken an Earth science course that included a major portion on astronomy. The students were asked to draw the Earth, the Sun, and the Moon. Nearly all of the students' drawings showed the Earth, the Sun, and the Moon to be about the same size or within a factor of 2 of each others' diameters—even though the Earth is 4 times the diameter of the Moon and the Sun is more than 100 times the diameter of the Earth (Driver et al. 1994).
- Mustafa (2007) interviewed students in Turkey who were approximately 14 years old, corresponding to ninth-grade students in the United States, concerning their ideas about the relative sizes of the Earth, Sun, and Moon. Of the 64 students interviewed, 28 (43%) knew that the Earth is smaller than the Sun and larger than the Moon; 10 of those who gave the correct answer (16% of the total number interviewed) could also say approximately how many times larger the Sun is than the Earth and the Moon. However, when the researcher asked the students to draw the Earth, the Sun, and the Moon, none of them attempted to show the correct scale; and when asked about the inconsis-

tency, they were not able to explain. The author concluded that "the students may have memorized the size of the Earth, but their alternative frameworks remained unchanged" (p. 46).

Suggestions for Instruction and Assessment

- Children in the primary grades can understand that if they hold up their thumb in front of a distant house, the thumb may completely cover the house but that does not mean their thumb is actually bigger than a house! However, extending that idea to astronomical objects like the Sun and the Moon takes a further leap, and requires prior understanding that Earth itself is a huge sphere that is so big that the part we see appears flat. So teaching about the relative sizes of the Earth, the Sun, and the Moon should not be attempted before upper elementary school (grades 4 and 5).
- Since textbook illustrations must greatly distort the sizes and distances of these bodies in order to fit them onto a single diagram, students are usually surprised to find out how different they are in size. Point out the difficulty of making scale size representations and why it is important to note when a representation does not represent actual scale size.
- Have students look at different textbooks and children's books that show representations of the Earth and the Sun in the same picture. Ask them to use the formative assessment classroom technique (FACT) representation analysis (Keeley 2008) to critique the picture, and ask them to explain what could be done to more accurately represent the actual difference in relative size between the Earth and Sun.
- Textbook illustrations showing distorted sizes tend to be remembered more readily than numbers, so it is not surprising that few middle school students know the vast difference in size between the Earth and the Sun. Relative size differences are more important to remember than absolute diameter measurements, which are hard to grasp because they are so large. Using examples like the pizza/toppings comparison will help your students remember the approximate size difference. Once students have experienced the pizza comparison, try other comparisons. For example, lay down a hula hoop and ask students to think of an object that would represent the Earth in relation to a hula hoop. Use increasing circle sizes to reinforce the notion of scale size comparison between the Sun and the Earth.
- High school students are expected to learn about the process of nuclear fusion that provides the Sun's energy. As a first step it is useful for students to know simple facts, such as how big the Sun is and what it is made from. One way to do this would be to ask the students to look up the diameter and volume of the Sun on the web and think of a way to communicate just how big it is to a middle school student. For example, how many Earths could line up across the Sun or how many Earths could fit inside the Sun?
- Extend the pizza scale size comparison to include the Moon and/or planets such as Mars and Jupiter.
- Once students have grasped the idea of scale size, have students compare relative distances based on the scale size of objects. See the next probe, "How Far Away Is the Sun?" as an example of combining scale size and distance.

References

American Association for the Advancement of Science (AAAS). 2009. Benchmarks for science literacy online. *www.project2061.org/publications/bsl/online*

Driver, R., A. Squires, P. Rushworth, and V. Wood-Robinson. 1994. *Making sense of secondary science: Research into children's ideas.* London: Routledge.

Jones, B. L., P. P. Lynch, and C. Reesink. 1987. Children's conceptions of the Earth, Sun, and Moon. *International Journal of Science Education* 9 (1): 43–53.

Keeley, P. 2008. *Science formative assessment: 75 practical strategies for linking assessment, instruction, and learning.* Thousand Oaks, CA: Corwin Press and Arlington, VA: NSTA Press.

Mustafa, C. 2007. Alternative views of the solar system among Turkish students. *Review of Education* 53: 39–53.

National Research Council (NRC). 1996. *National science education standards.* Washington, DC: National Academies Press.

Sadler, P. 1987. Misconceptions in astronomy. In *Second International Seminar on Misconception and Educational Strategies in Science and Mathematics*, ed. J. D. Novak, pp. 422–425. Ithaca, NY: Cornell University Press.

How Far Away Is the Sun?

Imagine a basketball represents the Sun. A seed, about 100 times smaller than the diameter of the basketball, represents the Earth. About how far away from the basketball should you place the "seed Earth" to show its distance from the Sun in this model? Circle the answer you think is closest to the relative distance between the "basketball Sun" and the "seed Earth."

A about 3 feet (or about 1 meter) away

B about 15 feet (or about 5 meters) away

C about 50 feet (or about 15 meters) away

D about 100 feet (or about 31 meters) away

E about 500 feet (or about 152 meters) away

F about 1,000 feet (or about 305 meters) away

Explain your thinking. Describe how you decided on your answer.

____

____

How Far Away Is the Sun?

Teacher Notes

Purpose

The purpose of this assessment probe is to elicit students' ideas about relative distance. The probe is designed to find out if students can apply the same scale used to represent the difference in sizes between the Earth and Sun to also represent the approximate distance between them.

Related Concepts

Apparent vs. actual size
Sun: distance, location relative to Earth

Explanation

The best answer is D: "About 100 feet (or 31 meters) away." More precisely, the ratio of the Sun's diameter at the equator to its average distance from Earth is 107.4. A very rough estimate can be made by taking the viewpoint of the seed that represents the Earth and backing away from the basketball until it appears to be about the same size as the Sun in the sky. Another way is to recall that the Sun is about 800,000 miles in diameter and 93,000,000 miles from Earth, a ratio of about 1:100.

Administering the Probe

This probe can be used with students in upper elementary grades through high school. It can be used to find out students' preconceptions related to the relative distance between the Sun and Earth, or it can be used following instruction about the solar system to see whether students developed a sense of scale size as a result of their instructional experiences. Make sure students know they are using the same scale to represent the sizes of the Earth and the Sun and the distance between them. To make the probe more interactive, use an actual basketball and a seed. Take the students into a long corridor, or go outside, and have them pace off in "feet" the distance they think represents the relative distance between the Earth and Sun at this scale and then commit to a written answer on the probe.

Related Ideas in *Benchmarks for Science Literacy* (AAAS 2009)

3–5 The Universe

- The Earth is one of several planets that orbit the Sun, and the Moon orbits around the Earth

Related Ideas in *National Science Education Standards* (NRC 1996)

5–8 Earth in the Solar System

★ The Earth is the third planet from the Sun in a system that includes the Moon, the Sun, eight other planets and their moons, and smaller objects such as asteroids and comets. *[Note: This standard was written before Pluto was reclassified.]* The Sun, an average star, is the central and largest body in the solar system.

Related Research

- Sadler (1992) developed a written test to measure high school students' understanding of astronomy concepts. An initial pilot of the test items with a small sample of students showed that approximately 80% of the students in grades 8–12 knew that the Earth is 93 million miles from the Sun, so a question asking how far Earth is from the Sun was thought to be too easy to use on the test. However, an item that was included in the test asked students to imagine a scale model in which the Sun is the size of a basketball, and to estimate how far from the basketball Earth should be located. The test was administered to 1,414 students in grades 8–12 who were just starting an Earth science or astronomy course. Responses were nearly equally distributed across choices of 5 feet, 10 feet, 25 feet, and 100 feet, indicating that memorizing the distance between the Earth and the Sun did not help students visualize the scale of the Earth-Sun system.
- Bakas and Mikropoulos (2003) gave a written questionnaire to 102 middle school students, ages 11–13, in Greece. They found that the majority of students (64%) understood that the Sun is much bigger than the Earth. However, only 16% were able to estimate the distance of the Earth from the Sun in a model in which the Sun is represented by an orange.
- Mustafa (2007) interviewed 64 students in Turkey about their understanding of the Earth, Moon, and Sun. The students were approximately 14 years old, corresponding to ninth-grade students in the United States. The students were asked: "If you used a basketball to represent the Sun, about how far away would you put a grape to represent the Earth?" Nineteen of the students (29%) had some knowledge of the Sun's distance; either they knew that the Sun is 300–400 times farther from the Earth than the Moon or they had correctly memorized the Sun's distance to be 149,600,000 kilometers from Earth. However, only one student correctly answered the question—that if the Sun is the size of a basketball, a grape that represents the Earth should be placed about 30 meters (98 feet) from the Sun.

Suggestions for Instruction and Assessment

- Elementary students should have opportunities to learn about scale models of things that they can see for themselves, such as cars, houses, and people. However, understanding scale models of the solar system

★ Indicates a strong match between the ideas elicited by the probe and a national standard's learning goal.

is much more challenging, and for most students this concept should be deferred to middle school.

- Middle school is an appropriate time to introduce students to scale models in which the same scale is used to represent both size and distance in the solar system. Although most middle school students have the ability to envision scale models, it is still best to begin with one idea at a time, such as the difference in size between the Earth and the Sun (e.g., using Probe 10, "Pizza Sun"), and then introduce the difference in distance at the same scale.
- The same approach can be used with the Earth and the Moon. For example, if Earth's diameter is represented by a basketball, the Moon, being about one-quarter the size of Earth, could be represented by a tennis ball. Most students are surprised that at this scale the Moon should be placed about 24 feet (7 meters) from the Earth.
- Since high school students are expected to learn about the relative scale of the solar system and our galaxy, it is important that they first have a basic understanding of the scale of the Sun-Earth system. They can be given the diameters of and distances between the Sun, the Earth, and the Moon and be challenged to choose a scale factor so as to create a scale model of the Earth-Sun-Moon system that will fit in the classroom (or gym or some other area of known size), using the same scale for size and distance. Many students may be able to memorize the numbers but will have difficulty envisioning a scale model until they actually work out the scaled distances and build the model for themselves.

References

American Association for the Advancement of Science (AAAS). 2009. Benchmarks for science literacy online. *www.project2061.org/publications/bsl/online*

Bakas, C., and T. A. Mikropoulos. 2003. Design of virtual environments for the comprehension of planetary phenomena based on students' ideas. *International Journal of Science Education* 25 (8): 949–967.

Mustafa, C. 2007. Alternative views of the solar system among Turkish students. *Review of Education* 53: 39–53.

National Research Council (NRC). 1996. *National science education standards.* Washington, DC: National Academies Press.

Sadler, P. M. 1992. The initial knowledge state of high school astronomy students. Doctoral diss., Graduate School of Education, Harvard University.

Sunspots

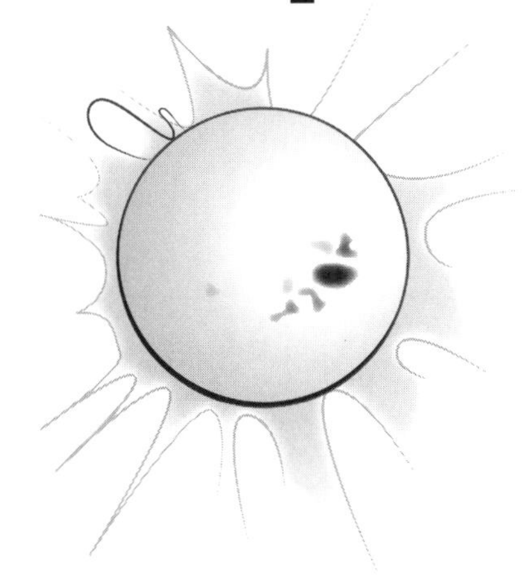

A group of students watched their teacher use a small mirror to project an image of the Sun in the classroom. The students thought it was really cool. The teacher told them that the few dark spots they saw were called sunspots. The students had a lot of different ideas about the sunspots. Here is what they said:

Jared: "If the Sun is so hot, maybe there are places that are a little cooler, so they look dark compared with the bright Sun."

Sharise: "Maybe sunspots are like islands. I read somewhere that the Sun is a huge 'sea' of hot glowing gas. But maybe there are 'islands' of dirt and rocks floating on the Sun."

John: "Spots on the Sun? I think there must have been dirt on the mirror or something. The Sun is a huge hot ball of gas, so it can't possibly have dark spots on it."

Latoya: "I think the spots you saw were clouds in our own atmosphere. If you watched the Sun for 15 minutes or so you would see them moving, and maybe new ones forming."

Emma: "I don't agree with any of you. I think there must be some other reason why there are spots on the Sun."

Which student do you agree with the most?_______________ Explain why you agree.

____

____

Sunspots

Teacher Notes

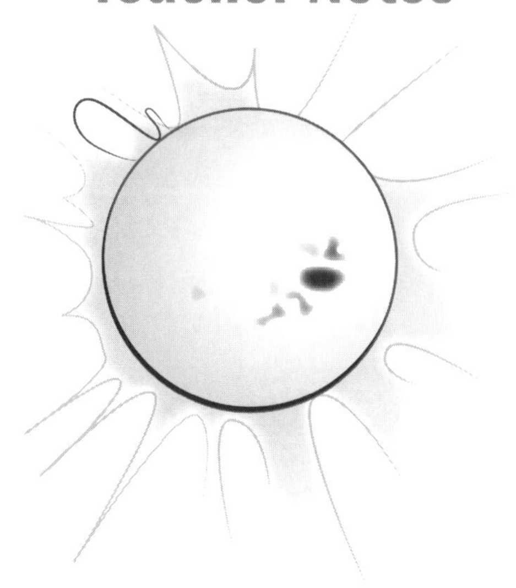

Purpose

The purpose of this assessment probe is to elicit students' ideas about sunspots. The probe is designed to determine if students are aware of a prominent feature of the Sun that they can observe under safe conditions—sunspots—and if they know what sunspots are.

Related Concepts

Solar system objects: spin
Sun: surface features

Explanation

Jared has the best answer: "If the Sun is so hot, maybe there are places that are a little cooler, so they look dark compared with the bright Sun." Sunspots can be seen, as described in the probe, and they change positions over several days as the Sun rotates. However, they are stable over short periods of time. Scientific research indicates that sunspots are areas of the Sun that are cooler than the surrounding regions—about 3,000–4,500°C versus 5,780°C.

Administering the Probe

This probe is best used with middle or high school students, depending on where in their curriculum students study the characteristics of objects in the solar system. The probe could be extended to ask students how they could test their ideas about sunspots.

Related Ideas in *Benchmarks for Science Literacy* (AAAS 2009)

6–8 The Universe

- The Sun is many thousands of times closer to the Earth than any other star.

Related Ideas in *National Science Education Standards* (NRC 1996)

K–4 Objects in the Sky

- The Sun, Moon, stars, clouds, birds, and airplanes all have properties, locations, and movements that can be observed and described.

5–8 Earth in the Solar System

★ The Sun, an average star, is the central and largest body in the solar system.
- The Sun is the major source of energy for phenomena on the Earth's surface, such as growth of plants, winds, ocean currents, and the water cycle.

Related Research

- Neil Comins's 2003 book *Heavenly Errors: Misconceptions About the Real Nature of the Universe* catalogs more than 1,700 misconceptions about astronomy gathered from his students at the University of Maine. Comins defines *misconceptions* as "deep-seated beliefs that are inconsistent with accepted scientific beliefs." In his book he explores several misconceptions in detail, and offers explanations for why they occur. The following are a few of the misconceptions about sunspots from his website, *www.physics.umaine.edu/ncomins:*
  - Sunspots are hotter places on Sun, not cooler.
  - Sunspots are permanent.
  - Sunspots are where meteors crash into (or are craters on) the Sun.
  - Sunspots are blemishes on photographs.
  - Sunspots are just "discolorations" to the Sun's surface.
  - Sunspots are regions of soot on the Sun.
  - Sunspots are volcanic in origin.
  - Sunspots are holes on the surface of the Sun.
  - Sunspots are places on the Sun that have run out of fuel to burn.
  - Sunspots are spots on the skin darkened by the Sun.

Suggestions for Instruction and Assessment

- **[Safety note: Observations of the Sun must always be preceded by safety instructions, since looking directly at the Sun, especially through a telescope, can cause blindness. However, the Sun can be safely observed as described in the probe, as a reflection on a wall inside a darkened room.]** Roll down blinds or curtains except for a small space at the bottom of one window. Hold a small mirror on the windowsill where the Sun is shining, and project an image of the Sun on the far wall or ceiling. Make sure students are all standing along the window side of the room, looking toward the image of the Sun projected on the opposite wall, so the Sun does not shine in their eyes. Although the image will be somewhat fuzzy, large sunspots will show up if there are any.
- Alternatively you can have the students view a real-time image of the Sun on the web in several different wavelengths at *http://sohowww.nascom.nasa.gov/data/realtime-images.html.*
- The students can learn about sunspots at these websites:
  - *www.exploratorium.edu/sunspots*
  - *www.universetoday.com/47728/what-are-sunspots*
  - *http://tuftsjournal.tufts.edu/2010/03_1/professor/01*

★ Indicates a strong match between the ideas elicited by the probe and a national standard's learning goal.

- The Sun is our closest star, so it is a very important object for students to learn about. Although sunspots are not typically identified as an essential topic in science standards, a study of sunspots over time makes an excellent science project and can help students recognize that the Sun is a real object in the solar system, with its own surface features and rate of rotation. A small telescope or binoculars with one lens blocked and held on a tripod can be used to project a sharp image of the Sun onto a piece of paper so sunspots can be traced or photographed, as described at *www.spaceweather.com/sunspots/doityourself.html.*
- Students can track the position of sunspots across the solar disc every day for a week and measure the rotation rate of the Sun (about 25 days at the equator to 36 days at the poles). Sunspotter telescopes are made by a variety of science supply companies and can be used to observe sunspots (see, e.g., *www.scientificsonline.com/Sunspotter.html).*

References

American Association for the Advancement of Science (AAAS). 2009. Benchmarks for science literacy online. *www.project2061.org/publications/bsl/online*

Comins, N. F. 2003. *Heavenly errors: Misconceptions about the real nature of the universe.* New York: Columbia University Press.

National Research Council (NRC). 1996. *National science education standards.* Washington, DC: National Academies Press.

Shorter Days in Winter

Mrs. Moro's students checked the newspapers every morning for the times of sunrise and sunset. They used this information to determine the number of hours of daylight. The class started this project in September, and by November they could see that the days were getting shorter and shorter. The students asked their families and neighbors to explain why days get shorter as winter approaches in the North. Here are the ideas they came to class with the next day:

Frank: "My mom says it's because of daylight saving time."

Jubal: "My sister said Earth's tilt causes the Sun to be farther away in winter."

Sybil: "My father thinks the angle of sunlight must be the cause."

Carter: "My brother says the Sun moves across the sky faster in winter."

Wendy: "My neighbor thinks the Sun's path in the sky gets shorter in winter."

Which student came to class with the best idea? _______________ Explain why you think that is the best idea.

____

____

Shorter Days in Winter

Teacher Notes

Purpose

The purpose of this probe is to elicit students' ideas about the changing length of daylight with the change in seasons. The probe is designed to find out if students can relate the apparent path of the Sun as seen from Earth to the length of daylight.

Related Concepts

Seasons: cause, length of day
Solar system objects: spin
Sun: path in the sky

Explanation

Wendy has the best answer: "My neighbor thinks the Sun's path in the sky gets shorter in winter." Her neighbor describes an accurate observation that the Sun makes a shorter and lower arc in the sky in winter. A shorter arc means the Sun spends less time above the horizon, so days are shorter. Here are problems with the answers given by the other students:

- Frank: "My mom says it's because of daylight saving time." Daylight saving time is a consequence of shorter days, not a cause.
- Jubal: "My sister said Earth's tilt causes the Sun to be farther away in winter." While it's true that Earth's axis is tilted, the tilt does not change the distance between the Earth and Sun. So that part of her sister's response is incorrect. However, the tilt of Earth's axis does cause the path of the Sun in the sky to appear to become shorter in winter and longer in summer. So this answer is partially correct.
- Sybil: "My father thinks the angle of sunlight must be the cause." It is true that as winter approaches not only does the path of the Sun grow shorter but also the Sun does not rise as high into the sky. The changing angle of the Sun results in cooler—but not shorter—days.
- Carter: "My brother says the Sun moves across the sky faster in the winter." Since

the apparent movement of the Sun across the sky is due to Earth's daily rotation on its axis, the rate of travel does not change.

Shorter days are one reason that it is colder in winter in the temperate and polar zones. When the Sun is above the horizon longer it has more time to heat up the land, and there is less time for the land to cool off at night. Also when the Sun is lower in the sky the energy we receive from the Sun is spread out over a larger area, so it is less intense, and the land absorbs less thermal energy.

Administering the Probe

This probe is best used at the middle and high school level. It can be combined with Probes 14, "Changing Constellations," and 15, "Why Is It Warmer in Summer?" to assess students' ideas about the seasons. If you are planning to use the probes to begin a unit of study, it is best to use this probe first, since it is important for your students to be able to explain why it is warmer in summer and cooler in winter from their own viewpoint, here on Earth, before taking the space view.

When you examine students' responses don't be surprised if most of them agree with Jubal, since the tilt of Earth's axis is related to a complete explanation of seasons. If that is the case facilitate a discussion in which students justify and argue their opinions.

Related Ideas in *Benchmarks for Science Literacy* (AAAS 2009)

K–2 The Universe

- The Sun, Moon, and stars all appear to move slowly across the sky.

3–5 The Earth

- The rotation of the Earth on its axis every 24 hours produces the night-and-day cycle. To people on Earth, this turning of the planet makes it seem as though the Sun, Moon, planets, and stars are orbiting the Earth once a day

6–8 The Earth

★ The number of hours of daylight and the intensity of the sunlight both vary in a predictable pattern that depends on how far north or south of the equator the place is. This variation explains why temperatures vary over the course of the year and at different locations.

9–12 The Earth

- Because the Earth turns daily on an axis that is tilted relative to the plane of the Earth's yearly orbit around the Sun, sunlight falls more intensely on different parts of the Earth during the year. The difference in intensity of sunlight and the resulting warming of the Earth's surface produces the seasonal variations in temperature.

Related Ideas in *National Science Education Standards* (NRC 1996)

5–8 Earth in the Solar System

★ Seasons result from variations in the amount of the Sun's energy hitting the surface, due to the tilt of the Earth's rotation on its axis and the length of day.

Related Research

- A review of 41 research studies on people's understanding of the seasons (Sneider, Bar, and Kavanagh 2011) found that

★ Indicates a strong match between the ideas elicited by the probe and a national standard's learning goal.

although understanding of seasons tended to increase with age, misconceptions were widespread even among college students and educated adults. For example, many people learn in school that seasons are due to Earth's tilt, but have the misconception that the tilt causes Earth to be closer to the Sun in summer. (In fact Earth is closest to the Sun in January, which is winter in the Northern Hemisphere.) The review article suggests a sequence of instruction beginning in elementary school, leading to full understanding of seasons in high school.

- Plummer (2008) interviewed 20 students in each of grades 1, 3, and 8. While she found a general trend toward higher levels of understanding among the older students, students at each grade level held misconceptions about how the Sun appeared to move through the sky during the day and how the Sun's path across the sky changed with the seasons. Many of the children at all ages thought that the Sun was directly overhead at noon every day, even though the Sun was never overhead at noon at the latitude where the children lived. Furthermore, there was no significant difference between third-grade students' and eighth-grade students' understanding of the Sun's apparent motions. However, she did find that students were able to learn about the Sun's changing path in the sky during the year with the help of a small planetarium.
- As part of an evaluation of a new high school course on astronomy, Sadler (1998) tested 1,250 high school students who had taken the course, which included instruction on the reasons for the seasons. One of the questions on the test was as follows: "The main reason for its being hotter in summer than winter is: (a) The Earth's distance from the Sun changes; (b) The Sun is higher in the sky; (c) The distance between the northern hemisphere and the Sun changes; (d) Ocean currents carry warm water north; and (e) An increase occurs in 'greenhouse' gases." Sadler found that most students who did poorly on the test overall chose the common misconception that Earth's distance from the Sun changes during the year. Most students who did moderately well chose a different misconception, that the distance between the Northern Hemisphere and the Sun changes, indicating that they understood that the tilt was somehow related to seasons; but they misunderstood *how* the tilt caused the seasonal variation in temperature. Only the students who did very well on the test overall chose the correct answer: "(b) The Sun is higher in the sky." Sadler concluded that these misconceptions should not be considered failures, but rather stepping-stones toward a full scientific understanding.

Suggestions for Instruction and Assessment

- One of the reasons it is so difficult for students to learn about the causes of Earth's seasons is that the space viewpoint is introduced too early, at the elementary level, before children have an opportunity to make the critical observations of the Sun's changing path in the sky that explain why it is warmer in summer and colder in winter.
- During the upper elementary grades students should have opportunities to observe the changing rising or setting point of the Sun as observed from a fixed location. September–October and March–April are good times to do that because the Sun's path changes most dramatically in the fall and spring. There are also a number of ways to safely measure the altitude of the Sun in the middle of the day, the simplest being to measure the length of a shadow of a fixed

object like a flagpole. Making these observations on a clear day once a week is sufficient to observe changes in the Sun's daily movements. It is also important for the teacher to help the students synthesize their observations to clarify the Sun's changing path in the sky and see that as winter approaches, the arcing path of the Sun becomes shorter, the sun does not travel as high in the sky, and the days grow shorter.

- A visit to a planetarium or use of a classroom planetarium can be very helpful in demonstrating these changes in the Sun's daily path, especially if it comes after students have made measurements of their own.
- Once students understand how the path of the Sun changes with the seasons, they should discuss how such changes affect average daily temperatures in the summer and winter. It should not be difficult for students in grades 4, 5, or 6 to understand that if the Sun is up longer, it will have more time to warm the Earth. That's why it's usually warmer in the afternoon, after the Sun has been up for a long time. It's also why it is coldest shortly before sunrise, because the land has been cooling off all night.
- Ideally this activity should take place after a unit on Earth's shape and gravity so that the students can understand that the Sun's apparent movement is due to Earth's daily spin and the speed of the Sun across the sky does not change. The length of the day is directly related to the length of the Sun's path in the sky.
- If middle school students are to fully understand the reasons for seasons, it is important that they first learn about the Sun's changing path in the sky during the year, before they are expected to explain these changes as a result of Earth's tilted axis.
- There are two reasons why the Sun's changing path in the sky causes it to be warmer in summer and cooler in winter. One reason concerns length of day. A more difficult reason to understand is that the altitude of the Sun in the sky makes a difference in the solar energy that reaches each square meter of ground. This can be addressed at the middle school level in a number of ways. Students can go outdoors to set up a solar cell that lies flat on the ground, connected to an ammeter. They can plot electrical current generated over a full day with no clouds and see that more energy is generated during the middle of the day. Or they can experiment in the lab with a bright incandescent lightbulb and solar cell or liquid crystal thermometer to see how angle of light affects the energy received.
- It should not be assumed that high school students understand the reasons for the seasons just because they studied it during elementary or middle school. This probe will tell you if high school students need to learn about the Sun's changing path in the sky. If they don't agree with Sybil's neighbor, it will be important that they also do activities like those discussed above, so that when they learn about Earth's tilted axis, they will understand that the model of the Earth's tilted axis and its orbit around the Sun is intended to explain what we experience on Earth. It is not simply an explanation to be memorized and repeated on a test.
- The National Science Digital Library's PRISMS (Phenomena and Representations for Instruction of Science in Middle Schools) website, funded by the National Science Foundation, has several representations and phenomena reviewed for their effectiveness in teaching ideas related to the seasons topic and subtopics: *http://prisms.mmsa.org.*

- Although the following Project 2061 AAAS website is still a work in progress, it has much useful information about students' ideas related to the seasons and visualizations that can be used to teach about the seasons: *http://flora.p2061.org/climate/#*.
- The GEMS (Great Explorations in Math and Science) guide, *The Real Reasons for Seasons,* has several activities that can be used to develop ideas related to this probe (Gould, Willard, and Pompea 2000).

References

American Association for the Advancement of Science (AAAS). 2009. Benchmarks for science literacy online. *www.project2061.org/publications/bsl/online*

Gould, A., C. Willard, and S. Pompea. 2000. *The real reasons for seasons: Sun-Earth connection.* GEMS. Berkeley, CA: Lawrence Hall of Science, University of California, Berkeley.

National Research Council (NRC). 1996. *National science education standards.* Washington, DC: National Academies Press.

Plummer, J. 2008. Students' development of astronomy concepts across time. *Astronomy Education Review* 7 (1): 139–148. *http://aer.aas.org/resource/1/aerscz/v7/i1/p139_s1*

Sadler, P. M. 1998. Psychometric models of student conceptions in science: Reconciling qualitative studies and distracter-driven assessment instruments. *Journal of Research in Science Teaching* 35 (3): 265–296.

Sneider, C., V. Bar, and C. Kavanagh. 2011. Learning about seasons: A guide for teachers and curriculum developers. *Astronomy Education Review* 10 (1). *http://aer.aas.org/resource/1/aerscz/v10/i1/p010103_s1*

Changing Constellations

Annie's class visited a planetarium, where they enjoyed a program called "Summer Constellations." During the program the instructor used her pointer to show the students how to find constellations in the summer sky. At the end of the program Annie asked if the instructor would point out Orion. The instructor said, "I can't do that because Orion isn't visible in the summer sky. Do you know why?"

Annie thought hard about why Orion might only be visible in the winter. Then, all of a sudden she had it! She stood up and explained why she thought Orion was not visible in the summertime, and the instructor said "Yes, that's absolutely correct!" Circle what you think Annie used to best explain her thinking.

A Earth's spin

B Earth's orbit

C Earth's spin and Earth's orbit

D Stars orbiting the Earth

Using the answer you selected above, provide an explanation for why Orion is not visible in the summer sky. ____

____

____

Changing Constellations

Teacher Notes

Purpose

The purpose of this assessment probe is to elicit students' ideas about why the constellations change with the seasons. The probe is designed to find out how students use the Earth's spin and orbit to explain why different constellations are visible in different seasons.

Related Concepts

Objects in the sky
Seasons: cause, constellations
Solar system objects: orbits, spin

Explanation

The best answer is C: Earth's spin and Earth's orbit. Annie may have said something like this: "We can only see about half of the constellations at any time because when we're facing the Sun it's daylight, so we can't see any stars. When our side of the Earth faces away from the Sun, it's nighttime so we can see the stars like you showed us here in the planetarium. But Earth is slowly going around the Sun; so six months later we won't be able to see these stars anymore because the Sun will be in the way, but we will be able to see the stars on the other side of the sky."

If the planetarium presenter is prepared, she will have a slide ready to illustrate Annie's explanation. It would look something like this.

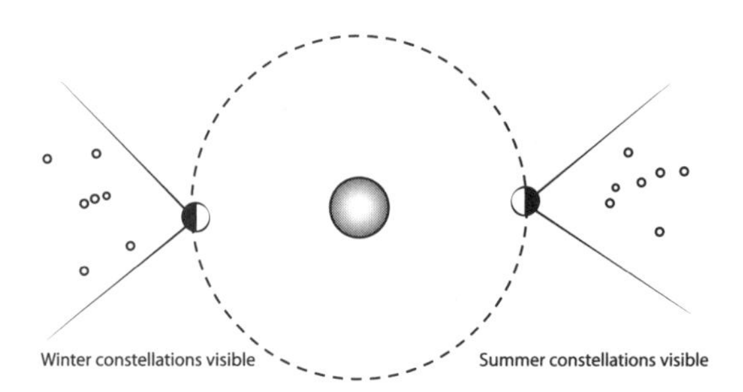

A student's ability to explain this phenomenon is an important step on the way to explaining other seasonal changes, such as the length of daylight and changes in weather conditions.

Administering the Probe

This probe is best used when introducing students to Earth's orbit around the Sun, which could be done as early as upper elementary school. It can also be used at the middle school and high school level as an introduction to a unit on the seasons to explain why we see different constellations during different seasons of the year. Make sure students understand that their written explanation should use the answer choice they selected as a starting point. They should then write what they thought Annie must have said before the planetarium instructor replied, "Yes, that's absolutely correct!" For students who wonder what Orion looks like, you may want to show a graphic of the constellation in the nighttime sky before administering this probe.

Related Ideas in *Benchmarks for Science Literacy* (AAAS 2009)

3–5 The Universe

- ★ The patterns of stars [constellations] in the sky stay the same, although they appear to move across the sky nightly, and different stars can be seen in different seasons.

Related Ideas in *National Science Education Standards* (NRC 1996)

5–8 Earth in the Solar System

- Most objects in the solar system are in regular and predictable motion. Those motions explain such phenomena as the day, the year, phases of the Moon, and eclipses.

Related Research

- Willard and Roseman (2007) relied on *Benchmarks* and findings from other studies to propose a learning progression for the seasons that required two essential prerequisites: (1) the link between sunlight and temperature and (2) an accurate understanding of how the Earth moves with respect to the Sun. Understanding why we see different constellations in different seasons can help students understand the Earth's annual motion around the Sun.
- A review of 41 research studies on people's understanding of the seasons (Sneider, Bar, and Kavanagh 2011) found that although understanding of seasons tended to increase with age, misconceptions were widespread even among college students and educated adults. Although many students struggle with the concept of seasons, a number of instructional approaches for middle school and high school students were found to be effective. One of the suggestions is to have students imagine what would happen if Earth had no tilt. Day and night would always be 12 hours long, and although there would be climate differences at different latitudes, there would be no variation in climate during the year. However, the phenomenon of different constellations during different seasons would be the same because that is a consequence of Earth's annual orbit around the Sun and is unrelated to tilt.

Suggestions for Instruction and Assessment

- Use this probe in conjunction with Probe 13, "Shorter Days in Winter," which focuses on the phenomenon of the changing length of the day during the year, and Probe 15, "Why Is It Warmer in Summer?"

★ Indicates a strong match between the ideas elicited by the probe and a national standard's learning goal.

during a unit on the seasons to find out how your students' ideas are developing. After the unit, you can give this sequence of probes again as a postassessment, asking the students to reconsider their responses.

- Bishop (2011) describes an activity that can be very effective in helping students understand why we see different constellations in different seasons. To prepare for this class, the teacher prepares 12 posters with the zodiacal constellations. Students stand in a large circle around the walls of the room holding these posters. One student stands in the center of the classroom, representing the Sun, and another student stands a few feet from the Sun, representing Earth in its orbit; the rest of the students watch from their seats. The student representing Earth is asked to turn away from the Sun and name the "constellations" she sees in the nighttime sky. Then she turns toward the Sun noting the constellations she *cannot* see because it is daytime. She then models what will occur six months later by walking halfway around to the other side of the Sun, where she is asked to do the same—observe zodiacal constellations in the nighttime sky and note which ones she *cannot* see when she faces the Sun because it is daytime. Another student can be asked to take the place of Earth and do the same to observe constellations in the fall and spring. Other students can take turns until all of the students can explain—in their own words—why we see different constellations in different seasons.
- The GEMS Teacher guide, *The Real Reasons for Seasons* (Gould, Willard, and Pompea 2000), describes a set of activities to address a variety of concepts (and misconceptions) related to the seasons.

References

American Association for the Advancement of Science (AAAS). 2009. Benchmarks for science literacy online. *www.project2061.org/publications/bsl/online*

Bishop, J. 2011. [Activity] F5. The Earth's revolution and the zodiac. In *Universe at your fingertips 2.0*, ed. A. Fraknoi. San Francisco: Astronomical Society of the Pacific. *www.astrosociety.org/uayf/index.html*

Gould, A., C. Willard, and S. Pompea. 2000. *The real reasons for seasons: Sun-Earth connection.* GEMS. Berkeley, CA: Lawrence Hall of Science, University of California, Berkeley.

National Research Council (NRC). 1996. *National science education standards.* Washington, DC: National Academies Press.

Sneider, C., V. Bar, and C. Kavanagh. 2011. Learning about seasons: A guide for teachers and curriculum developers. *Astronomy Education Review* 10 (1). *http://aer.aas.org/resource/1/aerscz/v10/i1/p010103_s1*

Willard, T., and J. E. Roseman. 2007. Progression of the reasons for seasons. Paper prepared for the Knowledge Sharing Institute of the Center for Curriculum Materials in Science, Washington, DC.

Why Is It Warmer in Summer?

Many textbooks say that Earth's tilt causes the change of seasons. But *how* does the tilt cause the seasons to change? Put an X next to any of the statements you think can help to explain how the tilt of the Earth causes it to be warmer in the summer than in the winter.

_______ **A** As the Earth circles the Sun, the direction of tilt relative to the plane of Earth's orbit gradually changes.

_______ **B** The direction of Earth's axis always stays the same as we circle the Sun.

_______ **C** When the Northern Hemisphere tilts toward the Sun we are closer to the Sun, so it is warmer.

_______ **D** When the Northern Hemisphere tilts toward the Sun the days are longer, so there is more time for the Earth to warm up.

_______ **E** When the Northern Hemisphere tilts toward the Sun then the Sun appears higher in the sky viewed from the United States, so sunlight is more concentrated and intense.

_______ **F** The Earth's tilt causes the Sun to be directly overhead at noon in the summer when viewed from the United States.

_______ **G** As the Earth circles the Sun it changes the angle of tilt during different seasons of the year, which then changes the amount of direct sunlight the Earth receives.

Explain your thinking on the back of this page. In your own words, describe how Earth's tilt relates to the change in seasons.

Why Is It Warmer in Summer?

Teacher Notes

Purpose

The purpose of this assessment probe is to elicit students' ideas about the cause of seasons. The probe is designed to find out how students use the Earth's tilted axis to explain seasonal changes.

Related Concepts

Seasons: cause, constellations, length of day
Solar system objects: orbits
Sun: altitude at noon, distance, path in the sky

Explanation

The best answers are B, D, and E. Each one of these statements provides a useful part of the answer to the question: Why is it warmer in the summer? The other statements are common misconceptions. The Earth's axis is tilted 23.4 degrees with respect to the plane of Earth's orbit around the Sun, and it maintains this direction throughout the year. (Note: The direction that the Earth's axis points does slowly change over very long periods of time during Earth's history in what is known as the precession of the equinoxes, but the change is negligible in a single year.) Because Earth's axis remains tilted in the same direction as it orbits the Sun, when one hemisphere is tilted toward the Sun, the other hemisphere is tilted away from the Sun. This has two effects:

1. When a hemisphere is tilted toward the Sun (e.g., Northern Hemisphere in June or Southern Hemisphere in December), light strikes parts of that hemisphere at a steeper angle, concentrating the Sun's energy and causing temperatures to get warmer. From the viewpoint of a person on Earth, the Sun climbs higher in the sky around midday. When a hemisphere is tilted away from the Sun (e.g., Northern Hemisphere in December or Southern Hemisphere in June), the sunlight is more spread out, concentrating less energy and resulting in cooler temperatures.
2. When a hemisphere is tilted toward the Sun, a larger portion of the hemisphere is

in full daylight. As viewed from a point on the surface, the Sun would appear to follow a longer path in the sky, so days are longer.

Both of these effects contribute to warmer summers and cooler winters.

Administering the Probe

This probe is best used at the middle and high school level. The justified list probe can be administered using the card sort strategy (Keeley 2008). Print each of the statements from the probe on cards and ask students to work in small groups to sort the cards into ones that can be used to explain why it is warmer in summer using ideas about Earth's tilt and ones that cannot be used to explain why it is warmer in summer using Earth's tilt. You may have a third category for those the group does not agree on or are unsure of at this time. As students sort the cards, they must defend their reasons for why they put it in each category.

Another recommended way to administer this probe is to collect the written responses and tally and post the results. Encourage discussion and see if any students wish to change their initial responses. Leading such a discussion will help you understand how your students are thinking about seasons. If you will be teaching a unit about the seasons, it is best to leave the tally in place until after the unit, and ask the students to reconsider their choices once more as a postassessment.

Related Ideas in *Benchmarks for Science Literacy* (AAAS 2009)

9–12 The Earth

★ Because the Earth turns daily on an axis that is tilted relative to the plane of the Earth's yearly orbit around the Sun, sunlight falls more intensely on different parts of the Earth during the year. The difference in intensity of sunlight and the resulting warming of the Earth's surface produces the seasonal variations in temperature.

Related Ideas in *National Science Education Standards* (NRC 1996)

5–8 Earth in the Solar System

★ Seasons result from variations in the amount of the Sun's energy hitting the surface, due to the tilt of the Earth's rotation on its axis and the length of the day.

Related Research

- A review of 41 research studies on people's understanding of the seasons found that although understanding of seasons tended to increase with age, misconceptions were widespread even among college students and adults. These studies suggested a number of insights into why it is so difficult to understand the reasons for the annual cycle of seasons (Sneider, Bar, and Kavanagh 2011).
- The most common misconception is that we are closer to the Sun in summer and farther away in winter. As students learned that the Earth's tilt was the cause of seasons, many students merged this new information with their prior misconception and developed a new misconception that the tilt made a significant difference in the distance of the two hemispheres from the Sun (Sneider, Bar, and Kavanagh 2011).
- A number of researchers suspected that common textbook illustrations of Earth's orbit around the Sun showing an elongated ellipse to indicate perspective misled

★ Indicates a strong match between the ideas elicited by the probe and a national standard's learning goal.

students about the changing distance from the Sun. This expectation turned out to be wrong, at least for ninth-grade students. However, the researcher noted that since some students could develop that misconception, teachers should point out that the elongated ellipse in the diagram is meant to show perspective, and that in fact the orbit is nearly circular (Lee 2010).

- Although many students struggle with the seasons concept, a number of instructional approaches for middle and high school students were found to be effective. These included having students observe changes in the path of the Sun during the year; teaching about seasonal changes at different latitudes, including the Southern Hemisphere; testing common misconceptions using physical models; and using 3-D modeling software (Sneider, Bar, and Kavanagh 2011).

Suggestions for Instruction and Assessment

- Learning the real reasons for seasons is a difficult task for most students because it involves an integration of several important concepts in astronomy, weather and climate, and the physics of light. Using this probe along with Probe 13, "Shorter Days in Winter," will reveal how students integrate these topics.
- This probe can also be combined with "Summer Talk" in *Uncovering Student Ideas in Science, Vol. 3: Another 25 Formative Assessment Probes* (Keeley, Eberle, and Dorsey 2008).
- Another piece of the seasons puzzle is how climate varies with latitude. Students need to understand that people in the Southern Hemisphere experience summer when we experience winter and vice versa. Also they should understand that climate varies very little with the seasons near the equator, but there are vast swings in seasonal climates near the poles. These differences are correlated with changes in the length of the day. A good way to involve students in learning about these differences is to assign students to use the internet to investigate how climate and day length change in different cities through the year.
- Yet another piece of the seasons puzzle, often overlooked, is that light travels in straight lines, and that light rays coming from the Sun to Earth are very nearly parallel. So, the energy in sunlight is more concentrated around the equator and is spread out closer to the poles. Avoid using the terms *direct* and *indirect* rays of the Sun, because these words tend to confuse students. Instead, refer to the angle at which light coming straight from the Sun strikes the ground.
- By the time students reach eighth or ninth grade, they should be ready to tie together the three major sets of ideas established in middle school and to develop a mental model of the Earth-Sun system to explain their Earth-based observations. A suggested sequence is as follows: In a darkened room with a brightly lit bulb in the middle, give small groups of students, standing around the bulb in a circle, an Earth globe to manipulate. Ask them to first consider how the length of day would vary if Earth's axis were not tilted. Identifying a point on the globe, students will observe that at every latitude there are 12 hours of daylight and 12 hours of night. Have the students move halfway around the circle and see that the hours of day and night do not change, as long as the axis is not tilted. Then have them make the same observation with the axis tilted; they will see that observers from the hemisphere tilted toward the Sun will experience a longer day and shorter night. Have them

move halfway around the circle again, keeping the tilt in the same direction, and they will see that the situation is now reversed, with observers in the Southern Hemisphere observing a shorter day and longer night. Do the same activity again, this time asking the students to imagine themselves standing on a point in the Northern Hemisphere and looking up at the Sun. How high in the sky do they need to look to see the Sun at noon? With some practice they should be able to observe that an observer would look higher in the sky to see the Sun when their hemisphere is pointed toward the Sun. Their days will also be longer at that time. They will also see that the opposite is true when they move their Earth globe to the other side of the Sun, provided that the tilted axis of the Earth continues to point in the same direction as it circles the Sun.

- You can test your students' success in merging the space view of the Earth and Sun with the observed path of the Sun in the sky by having them respond to the probe again after they have had an opportunity to develop formal conceptual understanding. After a discussion and final votes on what they think are the best answers to this probe, if the students still do not understand the concept, you can explain how a scientist would respond to the probe, using the opportunity to review the main points of the lesson. Administering the probe one more time at the end of the school year will allow you to find out how well your students retained what they learned after instruction.
- The National Science Digital Library's PRISMS (Phenomena and Representations for Instruction of Science in Middle Schools) website, funded by the National Science Foundation, has several representations and phenomena reviewed for their effectiveness in teaching ideas related to the seasons topic and subtopics*: http://prisms.mmsa.org.*

References

American Association for the Advancement of Science (AAAS). 1993. *Benchmarks for science literacy.* New York: Oxford University Press.

American Association for the Advancement of Science (AAAS). 2009. Benchmarks for science literacy online. *www.project2061.org/publications/bsl/online*

Keeley, P. 2008. *Science formative assessment: 75 practical strategies for linking assessment, instruction, and learning.* Thousand Oaks, CA: Corwin Press and Arlington, VA: NSTA Press.

Keeley, P., F. Eberle, and C. Dorsey. 2008. *Uncovering student ideas in science, vol. 3: Another 25 formative assessment probes.* Arlington, VA: NSTA Press.

Lee, V. R. 2010. How different variants of orbit diagrams influence student explanations of the seasons. *Science Education* 94 (6): 985–1007.

National Research Council (NRC). 1996. *National science education standards.* Washington, DC: National Academies Press.

National Research Council (NRC). 2011. *A framework for K–12 science education: Practices, crosscutting concepts, and core ideas.* Washington, DC: National Academies Press.

Sneider, C., V. Bar, and C. Kavanagh. 2011. Learning about seasons: A guide for teachers and curriculum developers. *Astronomy Education Review* 10 (1). *http://aer.aas.org/resource/1/aerscz/v10/i1/p010103_s1*

Section 3

Modeling the Moon

Concept Matrix: Modeling the Moon
Probes 16–27

PROBES	16. Seeing the Moon	17. Sizing Up the Moon	18. Does the Moon Orbit the Earth?	19. Earth or Moon Shadow?	20. Moon Phase and Solar Eclipse	21. Comparing Eclipses	22. Moon Spin	23. Chinese Moon	24. Crescent Moon	25. How Long Is a Day on the Moon?	26. Does the Earth Go Through Phases?	27. Is the Moon Falling?
GRADE-LEVEL USE →	K–12	K–5	3–12	5–12	6–12	8–12	6–12	5–12	K–12	6–12	6–12	6–12
RELATED CONCEPTS ↓												
apparent vs. actual size		X										
gravity			X									X
Moon: appearance	X	X		X	X	X	X	X	X			
Moon: eclipse					X	X						
Moon: orbit			X	X	X	X	X			X	X	X
Moon: phase				X	X				X		X	
Moon: size		X				X						
Moon: spin							X			X		
objects in the sky	X			X			X	X				
solar system objects: identity	X											
solar system objects: orbits			X		X					X	X	X
solar system objects: spin			X				X			X		
Sun: eclipse					X	X						

Teaching and Learning Considerations

Although it is extremely difficult for students to envision the Earth beneath their feet as a spherical body in space, it is much easier for them to picture the Moon as a spherical body illuminated by the Sun, circling the Earth and spinning on its own axis. That is why most of the probes in this book emphasize students' understanding of the Moon—because their mental models of the Moon will eventually transfer to their models of Earth in space, as just one among several planets in the solar system.

The experience of observing the Moon should begin as early as possible—kindergarten or first grade is recommended for the first two probes in this section. The Moon is fascinating to young children, and it is important to build on that fascination by encouraging them to observe the changing shape of the Moon and to see it in both the nighttime and daytime sky. Even young children can realize that the Moon is far bigger than it appears because it is very far way.

Probes 18–20 concern the Moon's orbit around Earth and the resultant phases as we see the Moon lit by the Sun at different locations in its orbit. These probes will enable you to see if your students make the common mistake of confusing the explanation of phases with the explanation for eclipses of the Moon. Probe 21, which asks why we see more eclipses of the Moon than eclipses of the Sun, will help you determine if your students have a clear mental model of the relationship between the Sun and the Moon as viewed from Earth.

Probes 22–26 challenge your students to envision a more complex mental model that involves all three bodies—Earth, Moon, and Sun—and to imagine what they would see from different perspectives, including from different locations on Earth, from the Moon, and from space, looking back at the Earth-Sun-Moon system.

The last probe in this section is an introduction to the next section, Dynamic Solar System, eliciting your students' understanding of orbits as they ponder the question of why the Moon doesn't fall to Earth.

Related Curriculum Topic Study Guides*

Earth, Moon, Sun System
Earth's Gravity
Motion of Planets, Moons, and Stars

*These guides are found in Keeley, P. 2005. *Science Curriculum Topic Study: Bridging the Gap Between Standards and Practice.* Thousand Oaks, CA: Corwin Press and Arlington, VA: NSTA Press. Each Curriculum Topic Study Guide provides a process to help the reader (1) identify adult content knowledge, (2) consider instructional implications, (3) identify concepts and specific ideas, (4) examine research on learning, (5) examine coherency and articulation, and (6) clarify state standards and district curriculum.

Related NSTA and Other Resources

NSTA Press Books

American Association for the Advancement of Science (AAAS). 2001. *Atlas of science literacy.* Vol. 1. (See "Gravity" map, pp. 42–43, and "Solar System" map, pp. 43–44.) Washington, DC: AAAS.

Ansberry, K., and E. Morgan. 2010. *Picture-perfect science lessons: Using children's books to guide inquiry, 3-6.* (See "The Changing Moon," pp. 247–262.) Arlington, VA: NSTA Press.

Gilbert, S. 2011. *Models-based science teaching.* Arlington, VA: NSTA Press.

Holt, G., and N. West. 2011. *Project Earth science: Astronomy.* 2nd ed. Arlington, VA: NSTA Press.

Keeley, P., F. Eberle, and L. Farrin. 2005. *Uncovering student ideas in science, vol. 1: 25 formative assessment probes.* (See "Gazing at the Moon," pp. 177–181, and "Going Through a Phase," pp. 183–187.) Arlington, VA: NSTA Press.

Keeley, P., F. Eberle, and J. Tugel. 2007. *Uncovering student ideas in science, vol. 2: 25 more formative assessment probes.* (See "Emmy's Moon and Stars," pp. 177–183, and "Objects in the Sky," pp. 185–190.) Arlington, VA: NSTA Press.

Keeley, P., and J. Tugel. 2009. *Uncovering student ideas in science, vol. 4: 25 new formative assessment probes.* (See "Moonlight," pp. 161–165, "Lunar Eclipse," pp. 167–171, and "Solar Eclipse," pp. 173–177.) Arlington, VA: NSTA Press.

Konicek-Moran, R. 2008. *Everyday science mysteries.* (See "Moon Tricks?" pp. 29–38.) Arlington, VA: NSTA Press.

Konicek-Moran, R. 2011. *Yet more everyday science mysteries.* (See "What's the Moon Like Around the World?" pp. 61–68.) Arlington, VA: NSTA Press.

NSTA Journal Articles

Bogan, D., and D. Wood. 1997. Simulating Sun, Moon, and Earth patterns. *Science Scope* 21 (2): 46, 48.

Brunsell, E., and J. Marcks. 2007. Teaching for conceptual change in space science. *Science Scope* 30 (9): 20–23.

Fidler, C., and C. Dotger. 2009. Visualizing the Earth and Moon relationship via scaled drawings. *Science Scope* 33 (4): 14–19.

Hall, C., and V. Sampson. 2009. Inquiry, argumentation, and the phases of the Moon: Helping students learn important concepts and practices. *Science Scope* 32 (8): 16–21.

Hermann, R., and B. Lewis. 2003. Moon misconceptions: Bringing pedagogical research of lunar phases into the classroom. *The Science Teacher* 70 (8): 51–55.

Hubbard, L. 2008. Bringing Moon phases down to earth. *Science and Children* 46 (1): 40–41.

Kruse, J., and J. Wilcox. 2009. Science sampler: Conceptualizing Moon phases—helping students learn how to learn. *Science Scope* 32 (5): 55–59.

Riddle, B. 2005. Science scope on the skies: You're blocking the light. *Science Scope* 29 (2): 70–72.

Riddle, B. 2006. Scope on the skies: The real shape of the Moon's orbit. *Science Scope* 30 (4) 68–70.

Riddle, B. 2007. Scope on the skies: Total lunar eclipse. *Science Scope* 30 (7): 76–78.

Rider, S. 2002. Perceptions about Moon phases. *Science Scope* 26 (3): 48–51.

Smith, W. 2003. Meeting the Moon from a global perspective. *Science Scope* 26 (8): 24–28.

Taylor, I. 1996. Illuminating lunar phases: Students construct fundamental knowledge of Moon phases. *The Science Teacher* 63 (8): 39–41.

Trundle, K., and T. Troland. 2005. The Moon in children's literature. *Science and Children* 43 (2): 40–43.

Trundle, K., S. Wilmore, and W. Smith. 2006. The Moon project. *Science and Children* 43 (6): 52–55.

Wallace, A., D. Dickerson, and S. Hopkins. 2007. Moon phase as a context for teaching scale factor. *Science Scope* 31 (4): 16–22.

Young, T., and M. Guy. 2008. The Moon's phases and the self shadow. *Science and Children* 46 (1): 30–35.

NSTA Learning Center Resources

NSTA SciGuides

http://learningcenter.nsta.org/products/sciguides.aspx

Earth and Sky (K–4 and 5–8)
Gravity and Orbits

NSTA SciPacks

http://learningcenter.nsta.org/products/scipacks.aspx

Earth, Sun, and Moon

NSTA Science Objects

http://learningcenter.nsta.org/products/science_objects.aspx

Earth, Sun, and Moon: Motion of the Moon

Solar System: The Earth in Space

Other Resources

Fraknoi, A., ed. 2011. *Universe at your fingertips 2.0.* San Francisco: Astronomical Society of the Pacific. Available at *www.astrosociety.org/uayf/index.html*

National Science Digital Library. PRISMS (Phenomena and Representations for Instruction of Science in Middle School) website: *http://prisms.mmsa.org*

Seeing the Moon

How often have you looked up into the sky and seen the Moon? Put an X next to all the times when you think you can go outside and see the Moon.

___ in the morning

___ at noon

___ in the middle of the afternoon

___ in the evening before sunset

___ in the evening after sunset

___ at midnight

Explain your thinking. How did you decide when you could see the Moon?

____

____

____

____

____

____

____

Seeing the Moon

Teacher Notes

Purpose

The purpose of this assessment probe is to elicit students' ideas about when we can see the Moon. The probe is designed to reveal whether students recognize that the Moon can be seen at different times during the daylight hours as well as at night.

Related Concepts

Moon: appearance
Objects in the sky
Solar system objects: identity

Explanation

The best answer is all of the choices. The Moon can be seen in the daytime as well as at night, although different Moon phases are seen at different times. The Moon can readily be observed during the evening hours, when a dark sky provides strong contrast to the bright Moon. What is less obvious and often goes unnoticed is that the Moon can also be seen during the day, although it is harder to see because there is less contrast between the Moon and the bright daytime sky.

There are other reasons that the Moon may not be visible. Either in daytime or nighttime the Moon may be below the horizon or obscured by clouds. During the new Moon phase, which lasts three or four days, the Moon is so close to the Sun in the sky that it cannot be observed in the daytime or nighttime.

The Moon can be observed in any of its phases during the day or night except for the full Moon, which is only visible at night or just at sunrise or sunset. That is because the full Moon is always opposite the Sun in the sky, so it is just rising when the Sun is setting, or just setting when the Sun is rising.

Administering the Probe

This probe can be used with students at all age levels. Although the concept is elementary, and people have many opportunities to see the Moon during the daytime, many adults have never noticed it except at night. Consequently

it is a good idea to administer this probe prior to any unit on the Moon, including high school, since students rarely take the time to observe the daytime and night sky. For the youngest students it's best to modify the probe by asking if the Moon can be seen only in the daytime, only at nighttime, or both. The probe can be extended for older students by asking them to draw a model to support their explanation.

Related Ideas in *Benchmarks for Science Literacy* (AAAS 2009)

K–2 The Universe

- ★ The Sun can be seen only in the daytime, but the Moon can be seen sometimes at night and sometimes during the day.
- The Moon looks a little different every day but looks the same again about every four weeks.

3–5 The Universe

- The Earth is one of several planets that orbit the Sun, and the Moon orbits around the Earth.

3–5 Constancy and Change

- Some things in nature have a repeating pattern, such as the day-night cycle, the phases of the Moon, and seasons.

Related Ideas in *National Science Education Standards* (NRC 1996)

K–4 Objects in the Sky

- ★ The Sun, Moon, stars, clouds, birds, and airplanes all have properties, locations, and movements that can be observed and described.

Related Research

- Children's early ideas about the Moon include the belief that the Moon is only visible at night or is in some way connected with the occurrence of night (Vosniadou and Brewer 1994).
- In a study by Sharp (1996) of 10- to 11-year-old students in England, 64% of the students did not believe that the Moon appears to move through the sky. Similar studies show that young students have not developed the understanding that celestial objects, such as the Moon, can be seen to move continuously, though very slowly.
- Many students believe that the Moon rises straight up, stays at the top of the sky throughout the night, and then sets straight down (Plummer 2009).
- A study of students in a small midwestern school revealed that 40% of the first graders (n = 20) believed that the Moon could only be seen at night, but by third grade 80% (n = 20) knew the Moon was visible during the day (Plummer and Krajcik 2010).

Suggestions for Instruction and Assessment

- Combine this probe with "Objects in the Sky" in *Uncovering Student Ideas in Science, Vol. 2: 25 More Formative Assessment Probes* (Keeley, Eberle, and Tugel 2007).
- It is not uncommon for children in the earliest grades to be taught the idea that "the Sun is for the day and the Moon is for the night," even though it is not true. While the Sun does indeed define daytime as the hours between sunrise and sunset, the Moon can be seen during the daytime or nighttime. That is why it is important that students have a chance to see the Moon during the day on occasion.
- Be aware that the misconception that the Moon is only visible at night may be

★ Indicates a strong match between the ideas elicited by the probe and a national standard's learning goal.

perpetuated by picture books young children see and read that associate the Moon with nighttime. Some books and nursery rhymes even depict the Moon as a character ready to go to sleep wearing a nightcap. Show children a picture book with one of these images and ask them if the Moon would ever come out in the day (Allen 2010).

- Understanding that the Moon can be seen both in the daytime and at night is a prerequisite to middle school students' identifying and explaining the pattern of moonrise and moonset times and providing evidence that the Moon must be slowly orbiting the Earth.
- You can check the time of moonrise in the nearest large city using a local newspaper or the internet. Then make sure that the Moon is not obscured by clouds before taking the children outdoors to see the Moon in the daytime sky.
- In the primary grades (K–2) students should be observing familiar objects in the sky, including the Sun and Moon, clouds, birds, and airplanes, noting which demonstrate regular patterns and which do not. Their observations of the daytime sky should include the changing positions of the Sun and Moon during the day, and changes in the Moon's apparent shape over about a month.
- In the upper elementary grades (3–5) students expand their observations and descriptions of the Sun and Moon to include stars and planets. They develop ideas about light reflection and light sources to explain why some things are seen in the daytime, some are seen at night, and others are seen in both the daytime and night. By fifth grade they begin to move from observations of the sky and describing patterns to developing explanations for these patterns.
- Have students spend a month making Moon observations, recording when they can see the Moon during the daytime and when they see the Moon at night. Have them record which Moon phase is visible at different times of the day as well as evening.
- Challenge older students to come up with an explanation and model for why a full Moon is not visible during the daytime.

References

Allen, M. 2010. *Misconceptions in primary science.* Berkshire, England: Open University Press.

American Association for the Advancement of Science (AAAS). 2009. Benchmarks for science literacy online. *www.project2061.org/publications/bsl/online*

Keeley, P., F. Eberle, and J. Tugel. 2007. *Uncovering student ideas in science, vol. 2: 25 more formative assessment probes.* Arlington, VA: NSTA Press.

National Research Council (NRC). 1996. *National science education standards.* Washington, DC: National Academies Press.

Plummer, J. 2009. Early elementary students' development of astronomy concepts in the planetarium. *Journal of Research in Science Teaching* 46 (2): 192–209.

Plummer, J., and J. Krajcik. 2010. Building a learning progression for celestial motion: Elementary levels from an Earth-based perspective. *Journal of Research in Science Teaching* 47 (7): 768–787.

Sharp, J. 1996. Children's astronomical beliefs: A preliminary study of year 6 children in Southwest England. *International Journal of Science Education* 18 (6): 685–712.

Vosniadou, S., and W. Brewer. 1994. Mental models of the day/night cycle. *Cognitive Science* 18: 123–183.

Sizing Up the Moon

Have you ever looked up at the Moon and wondered how big it is? Put an X next to the thing you think is closest to the size of the Moon.

_____ penny

_____ baseball

_____ basketball

_____ chair

_____ car

_____ my school

_____ my city

_____ Earth

_____ the Sun

Explain your thinking. How did you decide how big the Moon is?

____

____

____

____

____

____

Sizing Up the Moon

Teacher Notes

Purpose

The purpose of this assessment probe is to elicit students' ideas about the size of the Moon. The probe is designed to find out if students understand the idea that something huge can appear to be small if we see it from a great distance, and and to find out if they are able to apply that idea to realizing that the Moon is much bigger than it appears to be.

Related Concepts

Apparent vs. actual size
Moon: appearance, size

Explanation

The best answer is the Earth. Although the Moon may appear as small as a coin or baseball in the sky, that is because it is very far away. Things that look small in the night sky can actually be quite large. Of all the things on the list, the Moon is closest to the size of the Earth, although its diameter is only about 25% that of the Earth and only about 2% of the volume of the Earth.

Administering the Probe

This probe is best used with elementary students. For students in grades K–2 the probe can be presented verbally. For older students the probe can be extended by asking them to think about how they could measure the size of the Moon.

Related Ideas in *Benchmarks for Science Literacy* (AAAS 2009)

3–5 The Universe

- Stars are like the Sun, some being smaller and some larger, but so far away that they look like points of light.

3–5 Shapes

- Scale drawings show shapes and compare locations of things very different in size.

6–8 Shapes

- The scale chosen for a graph or drawing makes a big difference in how useful it is.

Related Ideas in *National Science Education Standards* (NRC 1996)

K–4 Objects in the Sky

- The Sun, Moon, stars, clouds, birds, and airplanes all have properties, locations, and movements that can be observed and described.

Related Research

- The age at which children are able to distinguish between how big something looks versus how big it really is has been of interest for more than half a century. The famous Swiss psychologist Jean Piaget reported that prior to about five years old few students are able to recognize that an object's actual size does not change when it is moved farther away. He called this ability "conservation of size," analogous to other conservation abilities (e.g., number, mass, liquid amount) that students develop as they mature (Piaget and Szeminska, 1952).
- Other psychologists confirmed Piaget's findings (Braine and Shanks 1965) and extended them. Rapoport (1967), for example, found that it wasn't until children were about nine years old that they could clearly and consciously distinguish between the actual and apparent size of an object.
- Tronick and Hershenson (1979) found that task difficulty also made a difference in whether or not a child was able to distinguish between apparent and actual size. For example, distinctions were easier when comparing a nearby object with an identical object just a few feet away, than when one object is much farther away (such as the Moon).
- Flavell (1986) confirmed earlier findings and conceived of the insight that what makes a difference in a child's ability to distinguish "real" from "apparent" is a theory of mind—that the child realizes that different people might see the same object in different ways. He noted that while some children could solve a simple task of this sort, it is not until they are 11 to 12 years old that they have a "substantial body of rich, readily available, and explicit knowledge in this area."
- This concept of judging relative size continues to be a rich area for research. Recently, Miller and Brewer (2010) studied the strategies that children in grades 1–3 used to judge the size of a disc at two different distances (6.1 meters and 61 meters). Some children were able to report the strategy they used; others could not. Students with higher visual spatial reasoning ability were more likely to be able to report on the strategies they used.

Suggestions for Instruction and Assessment

- In the primary grades (K–2) the challenge is for students to understand that things are not always as they seem. You might take the children outdoors to look at telephone poles and ask the students to compare how big the poles nearby *appear* compared with those farther down the street. The students can use their fingers held at arm's length to "measure" how big the different telephone poles appear to be. Then ask the students: "Do the closer telephone poles just look bigger, or are they really bigger?" This concept lays the groundwork for later understanding of the truly astronomical scale of the Moon and other planets. During the upper elementary grades (3–5), when students come to understand the Earth as an amazingly huge ball in space, they can begin to appreciate how large the Sun and Moon must be.

- Consider changing the context by asking students if the Sun is bigger than or smaller than the stars they see in the sky. This question reveals whether they recognize that a large object very far away can appear to be much smaller than a smaller object closer to the observer. The Sun, a medium-size star, is smaller than many large stars in the night sky.
- Consider extending this probe by asking upper elementary or middle school students how they would measure the size of the Moon. While elementary students may not have the mathematics knowledge of geometry to come up with the solution, the question elicits creative thinking. After middle school students have discussed their ideas about how to measure the size of the Moon, it is possible to provide some basic geometry activities where they can begin to learn how it was actually done in the days before space travel. *Project Earth Science: Astronomy* by Sean Smith (2001) suggests an activity called "It's Only a Paper Moon," in which students use similar triangles to find the diameter of the Moon, given its distance.
- In high school students have the mathematical background to figure out how to measure the size of the Moon. In order to measure the diameter of the Moon it is first necessary to measure its distance from Earth. Astronomy textbooks such as Project STAR (Coyle et al. 1993) provides several different ways to measure the Moon's distance (e.g., the parallax method, in which the Moon's position against background stars is observed from two different positions on Earth; and bouncing a radar pulse off the Moon and measuring the amount of time it takes for the beam to come back to the source).

References

American Association for the Advancement of Science (AAAS). 2009. Benchmarks for science literacy online. *www.project2061.org/publications/bsl/online*

Braine, M. D. S., and B. L. Shanks. 1965. The development of conservation of size. *Journal of Verbal Learning and Verbal Behavior* 4: 227–242.

Coyle, H., B. Gregory, W. M. Luzader, P. M. Sadler, and I. I. Shapiro. 1993. *Project STAR: The universe in your hands*. Dubuque, IA: Kendall Hunt.

Flavell, J. H. 1986. The development of children's knowledge about the appearance-reality distinction. *American Psychologist* 41 (4): 418–425.

Miller, B., and W. Brewer. 2010. Children's strategic compensation for size under constancy: Dependence on distance and relation to reasoning ability. *Visual Cognition* 18 (2): 296–319.

National Research Council (NRC). 1996. *National science education standards*. Washington, DC: National Academies Press.

Piaget, J., and A. Szeminska.1952. *The child's conception of number*. London: Routledge and Kegan Paul.

Rapoport, J. L. 1967. Attitude and size judgment in school age children. *Developmental Psychology* 38: 1187–1192.

Smith, S. 2001. *Project Earth science: Astronomy*. Arlington, VA: NSTA Press.

Tronick, E., and M. Hershenson. 1979. Size-distance perception in preschool children. *Journal of Experimental Child Psychology* 27: 166–184.

Does the Moon Orbit the Earth?

The Earth takes a year to orbit around the Sun. What about the Moon? Does it orbit the Earth? Circle the answer you think best describes the motion of the Moon.

A The Moon orbits the Earth about once a day.

B The Moon orbits the Earth about once a week.

C The Moon orbits the Earth about once a month.

D The Moon orbits the Earth about once a year.

E The Moon does not orbit the Earth.

Describe your thinking. Provide an explanation to support your answer.

___

___

___

___

___

Does the Moon Orbit the Earth?

Teacher Notes

Purpose
The purpose of this assessment probe is to elicit students' ideas about the Moon's motion. The probe is designed to find out whether students recognize that the Moon orbits the Earth, and, if so, how long they think it takes to complete one orbit.

Related Concepts
Gravity
Moon: orbit
Solar system objects: orbits, spin

Explanation
The best answer is C: "The Moon orbits the Earth about once a month." It is this monthly orbit that results in the cycle of lunar phases we see each month. As observed from Earth, it takes the moon approximately 27.3 days to go around the Earth once. This complete orbit is referred to as a sidereal month.

Administering the Probe
This probe is best used with upper elementary, middle school, or high school students, just before beginning lessons on Moon phases. It can also be used after students have had instruction in Moon phases to see how well students retain what they have been taught previously or whether they revert back to their preconceptions. For younger students or English-language learners, you can substitute *goes around* for the word *orbits.*

Related Ideas in *Benchmarks for Science Literacy* (AAAS 2009)

6–8 The Earth
★ The Moon's orbit around the Earth once in about 28 days changes what part of the Moon is lighted by the Sun and how much of that part can be seen from the Earth—the phases of the Moon.

Related Ideas in *National Science Education Standards* (NRC 1996)

5–8 Earth in the Solar System

★ Most objects in the solar system are in regular and predictable motion. Those motions explain such phenomena as the day, the year, phases of the Moon, and eclipses.

Related Research

- Sadler (1992) developed a written test to measure high school students' understanding of astronomy concepts, and gave it to 1,414 students in grades 8–12 who were just about to start a course in Earth science or astronomy. The test asked students to choose the best estimate of the time for the Moon to go around the Earth. The choices (and responses in percent) were (a) one hour (5%), (b) one day (37%), (c) one week (12%), (d) one month (37%), and (e) one year (7%).
- Schoon (1992) asked a similar multiple-choice question of 1,213 students in grades 5, 8, and 11; in college; and in trade school. The responses showed that 42% of the entire sample knew that the Moon takes one month to orbit Earth, 36% thought the Moon circled the Earth once a day, and 20% chose once a year. Although the correct answer, "one month," was somewhat more common among the older respondents, even one in four adults thought the Moon orbited the Earth once a day.

Suggestions for Instruction and Assessment

- Be aware that many students may recall from science class that the Earth goes around the Sun and the Moon goes around the Earth. However, these may be loosely remembered facts; it is more difficult to remember the details, such as how long it takes the Moon to circle the Earth, or how those facts explain phenomena such as night and day, Moon phases, and seasons.
- The best time to teach students about the Moon's orbit is when they are learning about Moon phases, so that the information becomes incorporated in their understanding of the Earth-Sun-Moon system. Research studies on children's learning about Moon phases suggest the best time to do that is in upper elementary or middle school, but no earlier than fifth grade.
- It is important for students to learn about the Moon's monthly cycle of phases by direct observation before teaching them the explanations for Moon phases and eclipses. While students in the primary grades can undertake observations of the Moon and create a "Moon calendar" to see the pattern of phase changes over a month, it is important to repeat these observations at the upper elementary or middle school level so that the observational facts of the Moon's cycle are fresh in students' minds before they are asked to model the phenomenon using light sources and balls to represent the Sun and Moon.
- Do not assume that most high school students know about the Moon's cycle of phases or that the phases are due to the Moon's monthly orbit around the Earth. Administer the probe first to learn more about your students' thinking before starting the unit.
- An advanced concept for high school students is that the Moon's monthly orbit with respect to the background stars is not the same as its orbit with respect to the Sun. Because the Earth advances about one-twelfth of the way around the Sun whereas the Moon circles the Earth once, the Moon

★ Indicates a strong match between the ideas elicited by the probe and a national standard's learning goal.

has to go a little farther to be in the same phase as on the previous cycle. The Moon's orbit around the Earth in relation to the Sun is called the Moon's synodic period, which is about 28.5 days. The Moon's orbit around the Earth with respect to the stars is called the Moon's sidereal period, which is about 27.3 days.

References

American Association for the Advancement of Science (AAAS). 2009. Benchmarks for science literacy online. Washington, DC: AAAS, Project 2061. *www.project2061.org/publications/bsl/online*

National Research Council (NRC). 1996. *National science education standards.* Washington, DC: National Academies Press.

Sadler, P. M. 1992. The initial knowledge state of high school astronomy students. Doctoral diss., Graduate School of Education, Harvard University.

Schoon, K. J. 1992. Students' alternative conceptions of Earth and space. *Journal of Geological Education* 40: 209–214.

Earth or Moon Shadow?

Two friends were looking at the Moon. Part of the Moon was visible to them. They wondered why they could only see part of the Moon. This is what they said:

Sally: "I think the part we can't see is the Moon's own shadow."

Enrique: "I think the Moon has moved into the Earth's shadow."

Circle the friend you agree with the most: Sally Enrique

Explain why you agree. ____

____

____

____

____

____

____

____

Earth or Moon Shadow?

Teacher Notes

Purpose

The purpose of this assessment probe is to elicit students' ideas about why the Moon appears to have different shapes at different times. The probe is designed to find out if students confuse the explanation for a lunar eclipse with the explanation for Moon phases.

Related Concepts

Moon: appearance, orbit, phase
Objects in the sky

Explanation

Sally has the best answer: "I think the part we can't see is the Moon's own shadow." Think of holding an orange next to a lamp in a darkened room: the part of the orange facing the lamp is brightly lit, but the part facing away from the lamp is dark. Enrique is expressing the common misconception that Moon phases are caused when the Moon enters the shadow of the Earth, which is an eclipse of the Moon. His erroneous view is supported by the observation that the shadow on a crescent Moon is curved, suggesting the Earth's shadow. However, at certain phases (first quarter and third quarter) there is a straight line between the light and dark parts of the Moon. If the Moon were entering Earth's shadow that line would always be curved due to the curved surface of the Earth.

Administering the Probe

This probe is best used with students in grades 5 and above. Make sure students know what is meant by "moving into Earth's shadow" (e.g., the Earth casts a shadow on the Moon) before asking students to respond. This probe is also appropriate for adults since many adults share Enrique's misconception.

Related Ideas in *Benchmarks for Science Literacy* (AAAS 2009)

K–2 The Universe

- The Moon looks a little different every day, but looks the same again about every four weeks.

6–8 The Earth

★ The Moon's orbit around the Earth once in about 28 days changes what part of the Moon is lighted by the Sun and how much of that part can be seen from the Earth—the phases of the Moon.

Related Ideas in *National Science Education Standards* (NRC 1996)

5–8 Earth in the Solar System

★ Most objects in the solar system are in regular and predictable motion. Those motions explain such phenomena as the day, the year, phases of the Moon, and eclipses.

Related Research

- A research review of 27 studies about children's and adults' understanding of Moon phases and eclipses (Kavanagh, Agan, and Sneider 2005) found that a very common misconception is that Moon phases occur when the Moon enters the Earth's shadow. That is a correct explanation for a lunar eclipse, but not for Moon phases. Findings from studies that attempted to teach students of various ages about the cause of Moon phases indicated that instruction is likely to be effective only for students in grades 5 and above. Furthermore, a large percentage of adults, including many teachers, are unable to separate the explanations for Moon phases and eclipses, suggesting that high school and college students can also benefit from instruction on this topic (Kavanagh, Agan, and Sneider 2005).
- Dai (1991) presented a unit on Moon phases to two fifth-grade classes in Taiwan in which students observed the Moon in the sky for several days and nights, used physical models of the Earth-Sun-Moon system to reproduce the effects of phases and eclipses, and acted out a play that demonstrated the movements of the Earth, Sun, and Moon. She compared the results of the instruction with the results in two other fifth-grade classes who learned about Moon phases using a traditional textbook approach, and she found the activity-based method to be more effective. Nonetheless, explaining Moon phases from the space viewpoint of the Earth-Sun-Moon system remained difficult for students even after instruction.
- Barnett and Morran (2002) conducted an in-depth study of 17 fifth-grade students as they learned about Moon phases and eclipses through class discussion, large- and small-group activities, and work with three-dimensional models. The researchers found that instruction was effective for many, though not all, students and recommended that rather than viewing misconceptions as impediments to learning, students' existing frameworks should be used to provide opportunities to reflect on their evolving understandings as they continue to learn science.

★ Indicates a strong match between the ideas elicited by the probe and a national standard's learning goal.

Suggestions for Instruction and Assessment

- This probe can be combined with "Going Through a Phase" in *Uncovering Student Ideas in Science, Vol. 1: 25 Formative Assessment Probes* (Keeley, Eberle, and Farrin 2005) and "Lunar Eclipse" in *Uncovering Student Ideas in Science, Vol. 4: 25 New Formative Assessment Probes* (Keeley and Tugel 2009).
- The Moon is fascinating to young children, so the elementary years are a good time to help them become familiar with the Moon's monthly cycle of phases. However, few children at this age level have sufficient spatial visualization abilities to understand the causes of Moon phases. And while the concept of an eclipse is easier to understand, until students fully appreciate the Earth as a spherical body in space, these concepts will not make a lot of sense.
- In the upper elementary grades (3–5), students should spend at least a month making Moon observations, recording when they can see the Moon during the daytime and when they see the Moon at night and noting how the relative positions of the Sun and Moon in the sky correspond to changes in the Moon's apparent shape.
- Researchers have found that students in fifth grade and above can begin to understand Moon phases, although many students may continue to confuse the explanations of Moon phases and eclipses, even after observing and modeling the Moon's monthly cycle (Kavanagh, Agan, and Sneider 2005).
- By middle school the great majority of students will have the spatial visualization skills needed to understand solar and lunar eclipses as well as phases. However, it is important that they first be clear about the monthly cycle of phases, preferably through their own observations, and then have an opportunity to model phases in order to explain why they occur.
- It is best to model both phases and eclipses with a single bright bulb in a darkened room to represent the Sun and a ball for each child to hold, representing his or her personal Moon. The students can then see their model Moon go through an entire cycle of phases as they slowly move it in a circle around their heads. The students can also observe when the Moon goes into the Earth's shadow (that is, the shadow of each student's own head) and distinguish that eclipse of the Moon from phases, which occur throughout the Moon's entire orbit.
- At the high school level, it is a good idea to present this probe before beginning a unit on astronomy to find out if your students understand the basic mechanics of the solar system. If they do not, the activity described above to model Moon phases and eclipses will be appropriate.
- The University of Nebraska–Lincoln has a good simulation to address the concept of lunar phases: *http://astro.unl.edu/naap/lps/animations/lps.html*

References

American Association for the Advancement of Science (AAAS). 2009. Benchmarks for science literacy online. *www.project2061.org/publications/bsl/online*

Barnett, M., and J. Morran. 2002. Addressing children's alternative frameworks of the Moon's phases and eclipses. *International Journal of Science Education* 24: 859–879.

Dai, M. 1991. Identification of misconceptions about the Moon held by fifth and sixth graders in Taiwan and an application of teaching strategies for conceptual change. Doctoral diss., University of Georgia.

Kavanagh, C., L. Agan, and C. Sneider. 2005. Learning about phases of the Moon and eclipses: A guide for teachers and curriculum

developers. *Astronomy Education Review* 4 (1): 19–52. *http://aer.aas.org/resource/1/aerscz/v4/i1/p19_s1*

Keeley, P., F. Eberle, and L. Farrin. 2005. *Uncovering student ideas in science, vol. 1: 25 formative assessment probes.* Arlington, VA: NSTA Press.

Keeley, P., and J. Tugel. 2009. *Uncovering student ideas in science, vol. 4: 25 new formative assessment probes.* Arlington, VA: NSTA Press.

National Research Council (NRC). 1996. *National science education standards.* Washington, DC: National Academies Press.

Moon Phase and Solar Eclipse

During a solar eclipse the Moon appears to completely cover the Sun. What phase is the Moon in just before and after a solar eclipse? Circle the answer that best matches your thinking.

A full Moon

B new Moon

C first quarter Moon

D last quarter Moon

E It can be in any phase.

Describe your thinking. Provide an explanation for your answer. ___________

____

____

____

____

Moon Phase and Solar Eclipse

Teacher Notes

Purpose

The purpose of this assessment probe is to elicit students' understanding of the Earth-Sun-Moon system well enough to explain the causes of Moon phases and solar eclipses and how these two phenomena are related.

Related Concepts

Moon: appearance, eclipse, orbit, phase
Solar system objects: orbits

Explanation

The best answer is B: new Moon. If students had an opportunity to observe and record Moon phases for a few weeks, they would see that Moon phase is correlated with the angle between the Moon and the Sun. Since the Moon must be between the Earth and the Sun for a solar eclipse to occur, the Sun must be illuminating the side of the Moon that we cannot see, which means it is in the new Moon phase. This idea is illustrated below (illustration is not to scale):

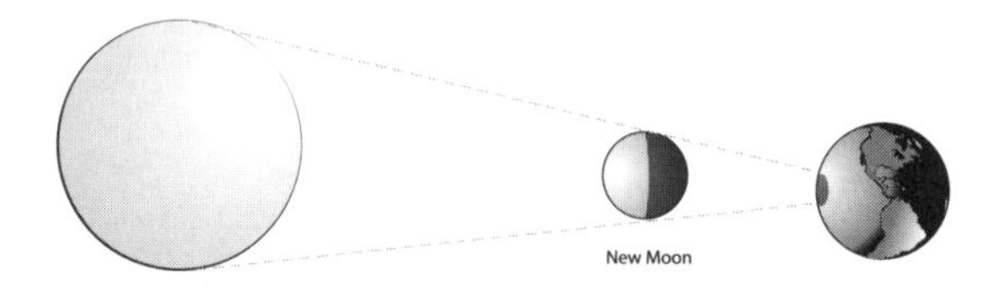

Administering the Probe

This probe is best used after middle or high school students have had the opportunity to learn about Moon phases and eclipses. The probe is a challenging question that requires students to stretch a bit to envision the relationship between Moon phases and eclipses. If necessary, show a graphic of each of the Moon phases so that the probe is not dependent on students knowing the terminology.

Related Ideas in *Benchmarks for Science Literacy* (AAAS 2009)

3–5 The Universe

- The Earth is one of several planets that orbit the Sun, and the Moon orbits around the Earth.

6–8 The Earth

- The Moon's orbit around the Earth once in about 28 days changes what part of the Moon is lighted by the Sun and how much of that part can be seen from the Earth—the phases of the Moon.

Related Ideas in *National Science Education Standards* (NRC 1996)

5–8 Earth in the Solar System

★ Most objects in the solar system are in regular and predictable motion. Those motions explain such phenomena as the day, the year, phases of the Moon, and eclipses.

Related Research

- Danaia and McKinnon (2007) administered tests of astronomy knowledge to 1,920 students in grades 7, 8, and 9 in Australia. One of the questions asked students for the phase of the Moon at a total solar eclipse. The correct answer, "new phase," was given by only 1% of the seventh graders and 10% of the eighth and ninth graders. About half of the students did not even attempt to answer the question.
- Several instructors of introductory college astronomy courses have used a standard test, the Astronomy Diagnostic Test (ADT), to measure the effectiveness of their teaching (Hufnagel 2001). For example, Zeilik and Morris (2003) used an early version of the ADT to evaluate a one-semester introductory course in astronomy for college freshmen at the University of New Mexico. One of the questions was: "When the Moon appears to completely cover the Sun (an eclipse), the Moon must be at which phase? (a) full; (b) new; (c) first quarter; (d) last quarter; (e) no particular phase." At the beginning of the course 49% knew the answer; at the end of the course 90% answered correctly.
- LoPresto (2006) used the ADT as a pre- and posttest for students in an introductory astronomy course at a community college. Test scores averaged over three years found (not surprisingly) that students tended to improve on items that were emphasized during the course. With respect to the question "When the Moon appears to completely cover the Sun (an eclipse), the Moon must be at which phase?" the percentage of students choosing the correct answer (new) increased from an average of 12% on the pretest to 42% on the posttest.

Suggestions for Instruction and Assessment

- This probe can be combined with "Solar Eclipse" from *Uncovering Student Ideas in Science, Vol. 4: 25 New Formative Assessment Probes* (Keeley and Tugel 2009).
- Students who have memorized a basic explanation for a solar eclipse (the Moon is between the Earth and the Sun) will probably have difficulty answering this question. However, those who learned to explain both phases and eclipses using a physical model have a good chance of envisioning how the Moon would be moving in its orbit just before passing in front of the Sun, and just after the eclipse.

★ Indicates a strong match between the ideas elicited by the probe and a national standard's learning goal.

- By middle school the great majority of students will have the spatial visualization skills needed to understand solar and lunar eclipses as well as phases. However, it will be important for them to first be clear about the monthly cycle of phases, preferably through their own observations, and then to have an opportunity to model phases.
- It is best to model both phases and eclipses with a single bright bulb in a darkened room to represent the Sun and a ball for each child to hold, representing his or her personal Moon. The students can then see their model Moon go through an entire cycle of phases as they slowly move it in a circle around their heads (with their heads representing Earth). The students will observe that as the Moon gets closer and closer to the Sun it has a thinner and thinner crescent. When it is very close to the Sun it cannot be seen at all—that is the new Moon phase. On rare occasions the Moon passes directly in front of the Sun, causing a solar eclipse—but only during the new Moon phase.
- This assessment probe is a sensitive instrument to determine not only if the students can recall a definition of a solar eclipse but also if they can envision the Moon getting closer and closer to the Sun, and entering the new Moon phase, before a solar eclipse occurs. Therefore, at the high school level, it may be a good idea to have students respond to this probe before beginning a unit on astronomy. If they have difficulties with it, the activity described above that models Moon phases and eclipses will be appropriate.

References

American Association for the Advancement of Science (AAAS). 2009. Benchmarks for science literacy online. *www.project2061.org/publications/bsl/online*

Danaia, L., and D. H. McKinnon. 2007. Common alternative astronomical conceptions encountered in junior secondary science classes: Why is this so? *Astronomy Education Review* 6 (2): 32–53. *http://aer.aas.org/resource/1/aerscz/v6/i2/p32_s1*

Hufnagel, B. 2001. Development of the Astronomy Diagnostic Test. *Astronomy Education Review* 1 (1): 47–51. *http://aer.aas.org/resource/1/aerscz/v1/i1/p47_s1*

Keeley, P., and J. Tugel. 2009. *Uncovering student ideas in science, vol. 4: 25 new formative assessment probes.* Arlington, VA: NSTA Press.

LoPresto, M. C. 2006. Astronomy Diagnostic Test results reflect course goals and show room for improvement. *Astronomy Education Review* 5 (2): 16–20. *http://aer.aas.org/resource/1/aerscz/v5/i2/p16_s1*

National Research Council (NRC). 1996. *National science education standards.* Washington, DC: National Academies Press.

Zeilik, M., and V. J. Morris. 2003. An examination of misconceptions in an astronomy course for science, mathematics, and engineering majors. *Astronomy Education Review* 1 (2): 101–119.

Comparing Eclipses

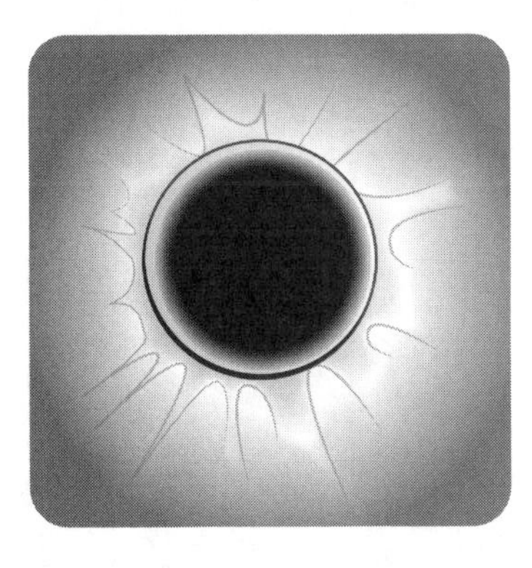

From any place on Earth a person can see more eclipses of the Moon than of the Sun. Why do you think this is so? Put an X in front of all the statements that support reasons why we see more lunar eclipses than solar eclipses.

_____ **A** The Sun moves more quickly than the Moon.

_____ **B** Anyone who can see the Moon when it enters Earth's shadow will see an eclipse of the Moon.

_____ **C** The shadow of the Moon on the Earth is very small and moves quickly.

_____ **D** The Moon goes in front of the Sun more often than the Sun goes in front of the Moon.

_____ **E** The Moon's orbit around the Earth is faster than Earth's orbit around the Sun.

_____ **F** The Moon spins on its axis faster than the Earth spins on its axis.

Describe your thinking. Use the ideas you marked with an X to explain why we see more lunar eclipses than solar eclipses. ______________________________

____

____

____

____

Comparing Eclipses

Teacher Notes

Purpose

The purpose of this assessment probe is to elicit students' ideas about eclipse phenomena. The probe is designed to determine if students have a mental model of the Earth-Sun-Moon system that allows them to figure out why we see more eclipses of the Moon than of the Sun.

Related Concepts

Moon: appearance, eclipse, orbit, size
Sun: eclipse

Explanation

The best answers are B and C. Both are needed to answer the question: Why do we see more eclipses of the Moon than of the Sun?

An eclipse of the Moon occurs when the Moon is in full phase, at the opposite side of the sky from the Sun. About twice per year the Moon moves into the shadow of the Earth, where it darkens and often turns a dark red color. It takes several hours until the Moon emerges from Earth's shadow. Anyone on the side of the Earth that can see the darkened Moon in Earth's shadow will observe an eclipse of the Moon. Since about two eclipses of the Moon occur every year, chances are 50% that we'll be on the right side of Earth to view the eclipse of the Moon, which is why a lunar eclipse can be seen from any spot on Earth about once a year.

An eclipse of the Sun occurs when the Moon passes between the Earth and Sun. Since the Moon is much smaller than the Earth, its shadow is also much smaller—only about 100 miles wide when it sweeps across the Earth. The only people who can see an eclipse of the Sun are those who are in the path of the Moon's shadow, which is why an eclipse of the Sun is seen only rarely from any given spot on the Earth.

Administering the Probe

This is a challenging probe, most appropriate for high school students, although middle school students who fully understand phases of the Sun and the Moon may be able to answer the probe's question. The probe would be most effectively used *after* a unit in which students learn about eclipses of the Sun and the Moon, to encourage them to envision and compare both types of eclipses. This probe can be used to help students construct and evaluate evidence necessary for scientific explanations. Encourage students to work in small groups to examine and critique each statement and decide whether or not it can be used to support an explanation for why we see more lunar eclipses than solar eclipses.

Related Ideas in *Benchmarks for Science Literacy* (AAAS 2009)

3–5 The Universe

- The Earth is one of several planets that orbit the Sun, and the Moon orbits around the Earth.

Related Ideas in *National Science Education Standards* (NRC 1996)

5–8 Earth in the Solar System

★ Most objects in the solar system are in regular and predictable motion. Those motions explain such phenomena as the day, the year, phases of the Moon, and eclipses.

Related Research

- Callison and Wright (1993) taught a unit on astronomy to college students preparing to become elementary teachers. To evaluate the effectiveness of their course, the researchers interviewed some of the students before and after instruction. They found that "conceptual growth occurs over time and requires revising and reflection." They also cautioned teachers to "beware of false positives," which are "responses that sound correct on the surface yet supported with incorrect notions. During the interview process the interviewer heard the 'right' words. However, once the subjects were questioned about the meaning of those 'right' words, the meaning was absent or incomplete." As an example, consider the following dialogue about Moon phases and eclipses between the interviewer (I) and a student (S):

I: What is your explanation for why the phases of the Moon occur?

S: Because the Moon is moving around the Earth and the Earth is rotating and the Moon gets light from the Sun. ... The phases occur because of what we can see from Earth of the Moon moving around the Earth.

I: If that is the case, would there be any time when the Earth was in between the Sun and the Moon? I would think it would cast a shadow on the Moon.

S: That would be an eclipse.

I: [In an attempt to get the subject to review her thinking]: So if the Earth is between the Moon and the Sun, what phase of the Moon do we see on Earth?

S: Dark.

I: So are you saying it is an eclipse?

★ Indicates a strong match between the ideas elicited by the probe and a national standard's learning goal.

S: Well, no, because an eclipse doesn't happen that often. I don't understand why an eclipse only happens every once in a while. It has just got to be on the correct axis or something. I don't know.

- Students' understanding of eclipse phenomena may be related to their lack of understanding of the relative sizes of the Sun, the Moon, and the Earth and the great distances between these bodies. Many students draw these objects so they are the same size or between half and double each other's size. They also draw the Sun and the Moon within one to four Earth diameters away from the Earth (Driver et al. 1994).

Suggestions for Instruction and Assessment

- This probe can be combined with the "Lunar Eclipse" and "Solar Eclipse" probes in *Uncovering Student Ideas in Science, Vol. 4: 25 New Formative Assessment Probes* (Keeley and Tugel 2009).
- Although middle school students can learn to explain Moon phases and eclipses of the Sun and the Moon, questions such as this one are very challenging and could perhaps be considered "stretch goals" for some students.
- The best way to explain why we see very few eclipses of the Sun is to have students stand in a circle around a single bright light in a darkened room and to hold a small ball representing the Moon at arm's length. The lightbulb represents the Sun and the students' heads represent the Earth. When the students hold the Moon ball in front of them so it covers the bright bulb, ask the students to glance around the room and see that everyone has a circular shadow over their eyes. Ask: Would everyone in the shadow see an eclipse of the Sun now? (Yes.) How about people not in the shadow, say those who "live" on your chin, cheek, or ear? (No.) Tell the students that the shadow of the real Sun on Earth during an eclipse of the Sun is just about 100 miles wide. So very few people get to see a solar eclipse, even though about two eclipses of the Sun occur every year. To see more eclipses it is necessary to travel to a place where the Moon's shadow will sweep across the Earth and hope that it will be a clear day.
- To explain why we can see about one eclipse of the Moon every year, have the students in the above demonstration move their Moon ball around to the full Moon position, so it is opposite the Sun bulb. Now have them pass the Moon into the shadow of their heads, noticing the shape of their head as they do so. Explain that at this point the Moon usually glows dull red and it is several hours before the Moon emerges from the shadow. Ask: Can everyone "living" on your face now see an eclipse of the Moon (Yes.) How about people "living" on the back of your head? (No. They cannot see the Moon now.) Explain that eclipses of the Moon occur about twice per year, and since there is a 50-50 chance of being on the right side of the Earth when it occurs, we usually have a chance to see an eclipse of the Moon about once a year.
- It is important for students to have a solid grasp of the Earth-Sun-Moon system by the time they reach high school and study more complex topics involving Earth's relationship to other bodies. This probe can be used to diagnose high school students' understanding of eclipses. If the students struggle with the answer, you may want to do the activity described above.

References

American Association for the Advancement of Science (AAAS). 2009. Benchmarks for science literacy online. *www.project2061.org/publications/bsl/online*

Callison, P. L., and E. L. Wright. 1993. The effect of teaching strategies using models on preservice elementary teachers' conceptions about Earth-Sun-Moon Relationships. Paper presented at the National Association for Research in Science Teaching (NARST), Atlanta, Georgia. (ERIC Document Reproduction no. ED360171)

Driver, R., A. Squires, P. Rushworth, and V. Wood-Robinson. 1994. *Making sense of secondary science: Research into children's ideas.* London: Routledge.

Keeley, P., and J. Tugel. 2009. *Uncovering student ideas in science, vol. 4: 25 new formative assessment probes.* Arlington, VA: NSTA Press.

National Research Council (NRC). 1996. *National science education standards.* Washington, DC: National Academies Press.

Moon Spin

People who observe the Moon notice that the same side of the Moon always faces the Earth. They also know that the Earth spins once a day on its axis. Does the Moon spin as well? If so, about how long does it take to make one full spin? Circle the answer that best matches your thinking.

A one hour

B one day

C one week

D one month

E one year

F Never. The Moon does not spin on its axis.

Explain your thinking. What evidence or model can you use to support your answer?

____

____

____

____

____

Moon Spin

Teacher Notes

Purpose

The purpose of this assessment probe is to elicit students' ideas about the Moon's motion. The probe is designed not only to find out if students know how long it takes for the Moon to make a complete rotation but also to uncover whether they think the Moon spins. Some students may cite evidence that the same side of the Moon always faces us to support their belief that the Moon does not spin.

Related Concepts

Moon: appearance, orbit, spin
Objects in the sky
Solar system objects: spin

Explanation

The best answer is D: one month. The Moon makes a complete spin on its axis about once every month. Its rotation (spin) is synchronous with its orbit around the Earth. Because of this we see the same side of the Moon all the time, which makes it appear as if the Moon is not spinning. Many people insist that the Moon does not turn or spin because we see the same face of the Moon throughout its lunar cycle.

Administering the Probe

This probe is best used at the middle or high school level, when students have well-developed spatial visualization abilities. At this level the probe can be an excellent elicitation to kick off a unit about the Moon, since students have to envision how the Moon orbits the Earth in order to answer this apparently simple yet challenging question.

Related Ideas in *Benchmarks for Science Literacy* (AAAS 2009)

6–8 The Earth

- The Moon's orbit around the Earth once in about 28 days changes what part of the Moon is lighted by the Sun and how much

of that part can be seen from the Earth—the phases of the Moon.

Related Ideas in *National Science Education Standards* (NRC 1996)

5–8 Earth in the Solar System

- Most objects in the solar system are in regular and predictable motion. Those motions explain such phenomena as the day, the year, phases of the Moon, and eclipses.

Related Research

- Kuethe (1963) administered a test of science questions to 100 college-bound high school graduates. When asked why we never see the other side of the Moon, 19 responded that it is because the Moon does not rotate.
- Sadler (1992) developed a written test for high school students that included the question: "Choose the best estimate of the time for the Moon to turn on its axis: (a) one hour; (b) one day; (c) one week; (d) one month; (e) one year." The test was administered to 1,414 students in grades 8–12 who were just about to start a course in Earth science or astronomy. The students' answers were approximately evenly divided among all answers, indicating that most of them guessed at the answer. Only 23% correctly responded that the Moon rotated once a month. Sadler noted that in light of the study by Kuethe (1963) showing that 19 out of 100 students thought the Moon did not rotate at all, in future versions of the test he would provide "does not rotate" in place of "one year."
- Dove (2002) reported the results of an end-of-year science exam given to 98 English students who were 12 years old and had recently completed lessons in astronomy. In response to a question about why we see only one face of the Moon, almost half of the students (46%) gave the correct answer—because the Moon rotates on its axis in the same time that it takes to orbit the Earth. The second most common answer was that the Moon does not rotate (14%), and the third most common response was that the Earth and Moon turn in unison (11%).

Suggestions for Instruction and Assessment

- An observation to be made during an upper elementary or middle school unit on the Moon is that we see only one "face" or side of the Moon, no matter what phase the Moon happens to be in. The famous "man in the Moon" is most apparent during the full Moon phase, so students may not notice the same features when they are only partially illuminated. However, students can look at images of the Moon taken through a telescope, showing the same major features at all phases.
- If students have completed a unit on Moon phases and eclipses and are ready to expand their learning, you can extend a Moon modeling activity by starting with this probe. Help the students figure out the correct answer on their own by modeling the Earth-Moon system. Have one student in the class play the part of the Earth and one person play the part of the Moon, and have the rest of the students imagine that they are sitting on the Sun watching. The Moon orbits the Earth by taking steps to the side, so the Moon always faces the Earth. From the Earth's perspective only the front of the person is visible, so it seems that the Moon is not spinning on its axis. But the rest of the students, watching from the Sun, can see the front and back of the

Moon—so they see it spin once on its axis at it orbits the Earth once.

- Do not assume that most high school students can understand this concept from reading about it in a textbook. Modeling the phenomenon as described above is essential for most people to understand that the Moon turns once on its axis as it orbits once around the Earth.

References

American Association for the Advancement of Science (AAAS). 2009. Benchmarks for science literacy online. *www.project2061.org/publications/bsl/online*

Dove, J. 2002. Does the man in the Moon ever sleep? An analysis of student answers about simple astronomical events: A case study. *International Journal of Science Education* 24 (8): 823–834.

Kuethe, J. L. 1963. Science concepts: A study of sophisticated errors. *Science Education* 47 (4): 361–364.

National Research Council (NRC). 1996. *National science education standards.* Washington, DC: National Academies Press.

Sadler, P. M. 1992. The initial knowledge state of high school astronomy students. Doctoral diss., Graduate School of Education, Harvard University.

Chinese Moon

Belinda lives in St. Louis, Missouri. She looked up at the sky one evening and observed a crescent Moon. What would her friend Lian, who lives in Beijing, China, see if she looked up at the Moon on the same night? Circle the shape of the Moon that best matches your thinking.

A crescent Moon

B quarter Moon

C gibbous Moon

D full Moon

E new Moon

Explain your thinking. What rule or reasoning did you use to select your answer?

____

____

____

____

____

____

Chinese Moon

Teacher Notes

Purpose

The purpose of this assessment probe is to elicit students' ideas about Moon phases when observed from different locations. The probe is designed to determine if students are able to take the "space point of view" when thinking about Moon phases and realize that the Moon will be in the same phase, no matter from where on Earth it will be viewed.

Related Concepts

Moon: appearance
Objects in the sky

Explanation

The correct answer is A. Lian would see a crescent Moon—the same Moon phase that Belinda sees. When imagining how the Earth-Sun-Moon system would appear from space, it is evident that the Moon's position in its orbit around Earth determines the phase. Since Lian lives in China, which is several time zones away from where Belinda lives, she will see it at a different time; but the Moon does not move very far in its month-long orbit, so the phase would be the same for both observers.

Administering the Probe

This probe is best used with middle and high school students. It is a good probe to use before, during, or after a lesson on Moon phases—or at all three times! If students are unfamiliar with the names of the general Moon phase shapes, show them a picture of each Moon phase shape.

Related Ideas in *Benchmarks for Science Literacy* (AAAS 2009)

K–2 The Universe

★ The Moon looks a little different every day but looks the same again about every four weeks.

6–8 The Earth

★ The Moon's orbit around the Earth once in about 28 days changes what part of the Moon is lighted by the Sun and how much of that part can be seen from the Earth—the phases of the Moon.

Related Ideas in *National Science Education Standards* (NRC 1996)

5–8 Earth in the Solar System

★ Most objects in the solar system are in regular and predictable motion. Those motions explain such phenomena as the day, the year, phases of the Moon, and eclipses.

Related Research

- Schoon (1992) gave a written questionnaire to 1,213 elementary, secondary, and adult students in which they were asked if people in Australia and the United States saw the same phase of the Moon on the same night. More than half (52%) chose the misconception that people would see a different phase of the Moon. Although older students tended to do better, the differences within the age groups were greater than differences between groups. After the students answered the questions, the researcher led a discussion. Students reported that they found the problem to be very difficult, and most used one of two means to solve it: either imagining themselves in space looking back at the Earth and Moon, or recalling that Moon phases are sometimes listed on calendars, suggesting that phase is independent of location.
- Rider (2002) interviewed 32 middle school students representing a range of ability, effort, and academic success in science who had recently learned about the Moon in school. When the students were asked: "If there is a full Moon today in New York State, will there be a full Moon today in California?" only about 10% of the students recognized that observers in both locations would see the same Moon phase on the same day.
- Subramaniam and Padalkar (2009) conducted a detailed study of conceptual change related to Moon phases among eight graduate students, and found that "a major source of difficulty for the participants was to understand that the Earth's rotation or the observer's position on the Earth has no causal role in the occurrence of the lunar phases, but only determines when and whether the Moon is visible at all" (Subramaniam and Padalkar 2009, p. 409).

Suggestions for Instruction and Assessment

- This probe can be combined with "Gazing at the Moon" in *Uncovering Student Ideas in Science, Vol. 1: 25 Formative Assessment Probes* (Keeley, Eberle, and Farrin 2005).
- Thinking about this question requires that the students envision where the Moon must be in its orbit for Belinda to see what she sees. If they can do that, it is a relatively small step to envision what Lian would see.
- Since middle school is often the time when students develop explanations for Moon phases, they need opportunities to model

★ Indicates a strong match between the ideas elicited by the probe and a national standard's learning goal.

Moon phases. Place a single bright bulb in the middle of a darkened room to represent the Sun, and have the students stand in a large circle around the lit bulb. Give each child a ball to hold, representing his or her personal Moon. They can then see the model Moon go through an entire cycle of phases as they slowly move it in a circle around their head. Tell the students that their head represents the Earth. Ask them to imagine that their right eye represents New York City and their left eye represents Beijing, China. Ask them to close first one eye, then the other to see if they see different phases or the same phase. (They will see the same phase.)

- Once students understand Moon phases by modeling, show them a diagram like that often presented in textbooks about Moon phases to develop conceptual understanding. Ask them to notice where the Moon must be at first quarter phase and to imagine a line from different parts of the Earth to the Moon in that part of its orbit. Does everyone on Earth see the same phase? (Since such drawings are not to scale, the students will notice that there is a slight difference in phase. Be sure to acknowledge this excellent observation, but also point out that the phase is not terribly different, such as between a crescent and full Moon.)
- As a stretch goal, challenge students to envision how the Moon would appear on the same night viewed from New York City and Sydney, Australia. Although the Moon phase would be the same when viewed from the Northern and Southern hemispheres, a crescent Moon would appear reversed in the sky, with the "points" of the crescent on the right rather than the left side. That is because a person standing in the Southern Hemisphere would be "upside down" compared with a person standing in the Northern Hemisphere. To see what someone in Australia would see, you would need to stand on your head while looking at the Moon.
- Although the explanation for Moon phases is typically an upper elementary or middle school learning objective, high school students often need a refresher lesson. This probe can provide an excellent focus for such a lesson. If discussion is not sufficient for the students to understand the concept, then it is best to go through the activity mentioned above for middle school students; in this way they are likely to learn it at a much deeper level than from a lecture or discussion alone.

References

American Association for the Advancement of Science (AAAS). 2009. Benchmarks for science literacy online. *www.project2061.org/publications/bsl/online*

Keeley, P., F. Eberle, and L. Farrin. 2005. *Uncovering student ideas in science, vol. 1: 25 formative assessment probes.* Arlington, VA: NSTA Press.

National Research Council (NRC). 1996. *National science education standards.* Washington, DC: National Academies Press.

Rider, S. 2002. Perceptions about Moon phases. *Science Scope* 26 (3): 48–51.

Schoon, K. J. 1992. Students' alternative conceptions of Earth and space. *Journal of Geological Education* 40: 209–214.

Subramaniam, K., and S. Padalkar. 2009. Visualization and reasoning in explaining the phases of the Moon. *International Journal of Science Education* 31 (3): 395–417.

Crescent Moon

When there is a crescent Moon in the night sky, how much of the *entire* Moon's spherical surface is actually lit by the Sun? Circle the answer that best matches your thinking.

A quarter or less of the entire Moon

B half of the entire Moon

C three quarters of the entire Moon

D the entire Moon

Explain your thinking. Provide an explanation for your answer. ______________

____

____

____

____

____

Crescent Moon

Teacher Notes

Purpose

The purpose of this assessment probe is to elicit students' ideas about how much of the spherical Moon is lit by the Sun at any point in time. The probe is designed to reveal whether students recognize that at any point in time the Sun shines on half of the Moon, even though we see only a part of the lit face of the spherical Moon during the crescent, quarter, or gibbous phase.

Related Concepts

Moon: appearance, phase

Explanation

The best answer is B: "half of the entire Moon." Like sunlight hitting the Earth, the Sun also illuminates one half of the Moon (except for the occasion for a few hours twice a year, when the Moon enters the shadow of Earth). Moon phases occur because we see only a portion of the lit-up half of the Moon as it orbits the Earth. Only during the full Moon phase do we see the entire lit portion of the Moon.

Administering the Probe

This probe can be used at all grade levels, once students understand that we see the Moon by its reflected light. Even though the probe is asking how much of the Moon is actually lit by the Sun (not how much you see that is lit by the Sun), students are likely to choose the response that matches what they actually *see* as opposed to the entire spherical moon, because they may never have thought before about how much of the entire Moon is actually lit by the Sun. Some students may think that the "entire Moon" means the entire lit face of the Moon, so it is important to ask them to explain why they chose a given answer (without cuing them that they need to think spherically) and to encourage discussion afterward that reveals the difference between what they actually see facing them versus the entire moon.

Related Ideas in *Benchmarks for Science Literacy* (AAAS 2009)

K–2 The Universe

- The Moon looks a little different every day but looks the same again about every four weeks.

3–5 The Universe

- The Earth is one of several planets that orbit the Sun, and the Moon orbits around the Earth.

6–8 The Earth

★ The Moon's orbit around the Earth once in about 28 days changes what part of the Moon is lighted by the Sun and how much of that part can be seen from the Earth—the phases of the Moon.

Related Ideas in *National Science Education Standards* (NRC 1996)

K–4 Changes in Earth and Sky

- The observable shape of the Moon changes from day to day in a cycle that lasts about a month.

5–8 Earth in the Solar System

- Most objects in the solar system are in regular and predictable motion. Those motions explain such phenomena as the day, the year, phases of the Moon, and eclipses.

Related Research

- Trundle, Atwood, and Christopher (2007) investigated the learning process of 12 preservice elementary teachers concerning Moon phases and eclipses. They found that the realization that the Moon is always half illuminated by the Sun (except for the rare occurrence when the Moon moves into Earth's shadow) was an important step in the teachers' ability to fully understand Moon phases.
- Subramaniam and Padalkar (2009) used visualization techniques to help graduate students learn about Moon phases. One method was to ask the teachers to visualize a person's head illuminated by a single light, as a model for the Moon. They found that in order to succeed at this task, it was important for the learners to realize that half of the spherical Moon's surface is always lit and for them to focus on the shape of the boundary between light and dark.

Suggestions for Instruction and Assessment

- Combine this probe with "Moonlight" in *Uncovering Student Ideas in Science, Vol. 4: 25 New Formative Assessment Probes* (Keeley and Tugel 2009).
- Demonstrate how a sphere is illuminated in a dark room. Have students stand where they see half of the sphere lit up by the light source. Then have them change their position in relation to the light and the sphere to see how their view of the lit part of the sphere changes depending on where they are standing, resulting in different phases—crescent, quarter, gibbous, and full. As one student walks around the sphere, have another student stand where half of the sphere is fully illuminated and say what they see. This modeling activity should help your students see that half the Moon is lit all the time and that we only see a part of what is lit, depending on our position in relation to the Sun, Moon, and Earth.

★ Indicates a strong match between the ideas elicited by the probe and a national standard's learning goal.

References

American Association for the Advancement of Science (AAAS). 2009. Benchmarks for science literacy online. *www.project2061.org/publications/bsl/online*

Keeley, P., and J. Tugel. 2009. *Uncovering student ideas in science, vol. 4: 25 new formative assessment probes.* Arlington, VA: NSTA Press.

National Research Council (NRC). 1996. *National science education standards.* Washington, DC: National Academies Press.

Subramaniam, K., and S. Padalkar. 2009. Visualization and reasoning in explaining the phases of the Moon. *International Journal of Science Education* 31 (3): 395–417.

Trundle, K. C., R. K. Atwood, and J. E. Christopher. 2007. A longitudinal study of conceptual change: Preservice elementary teachers' conceptions of Moon phases, *Journal of Research on Science Teaching* 44 (2): 303–326.

How Long Is a Day on the Moon?

Four students were designing a Moon base for a science project. Planning the Moon base was easy. But deciding what a day-night cycle on the Moon base would be like was hard! All four students had different ideas. Here is what they said:

Hannah: "I think the length of the day-night cycle on the Moon is 24 hours."

Sachet: "It depends where the Moon base is. If it is on the dark side of the Moon, there will never be daytime."

Ravi: "I think there would be about two weeks of sunlight and two weeks of darkness."

Manuel: "It depends on the Moon phase. In a crescent Moon, daylight would be much shorter. When there's a full Moon, daylight would be much longer."

Which student do you think has the best idea? ____________________ Explain why you agree.

____

____

____

____

How Long Is a Day on the Moon?

Teacher Notes

Purpose

The purpose of this assessment probe is to elicit students' ideas about the day-night cycle on the Moon. The probe is designed to find out if students can coordinate two concepts: the Moon turns once on its axis each month, and half of the Moon is always in sunlight.

Related Concepts

Moon: orbit, spin
Solar system objects: orbits, spin

Explanation

Ravi has the best answer: "I think there would be about two weeks of sunlight and two weeks of darkness." Half of the Moon is always in sunlight and half in darkness. The Moon turns slowly on its axis, making one full turn in the same time it takes to orbit the Earth, which takes about four weeks or one month (more precisely 27.3 days). So a "full day" on the Moon would last a month, with two weeks of sunlight and two weeks of darkness.

Administering the Probe

This is a challenging question, most appropriate for high school students, although middle school students who fully understand the Moon-Sun relationship may be able to answer it. The probe would be most effectively used *after* a unit in which students learn about Moon motion, to encourage them to envision the Moon from this perspective. This probe can be used to help students construct and evaluate evidence necessary for scientific explanations. Encourage students to work in small groups to examine and critique each statement and decide whether or not it can be used to support an explanation for the length of a Moon day.

Related Ideas in *Benchmarks for Science Literacy* (AAAS 2009)

6–8 The Earth

- ★ The Moon's orbit around the Earth once in about 28 days changes what part of the Moon is lighted by the Sun and how much of that part can be seen from the Earth—the phases of the Moon.

Related Ideas in *National Science Education Standards* (NRC 1996)

5–8 Earth in the Solar System

- Most objects in the solar system are in regular and predictable motion. Those motions explain such phenomena as the day, the year, phases of the Moon, and eclipses.

Related Research

- Dove (2002) asked 98 12-year-olds how long a full day and night would be at a space station on the Moon. Although 46% of the students knew that we only see one face of the Moon because it rotates during the same period of time that it makes a complete orbit of Earth, only 39% knew that the period of time between two sunrises at a space station on the Moon would be about one month.
- Trundle, Atwood, and Christopher (2002) conducted an experiment with 63 college students who were studying to be elementary science teachers. At the start of the experiment, less than 10% of the college students understood the scientific explanation for Moon phases, which included the idea that half of the Moon is always in sunlight. Instruction involved small-group work during which the students observed the Moon in the sky over a period of nine weeks, used physical models with guidance from the instructor, and discussed their evolving ideas with each other. Three weeks after instruction, 68% of the experimental group had a full understanding of Moon phases, while 94% had at least some understanding of the scientific explanation for Moon phases.

Suggestions for Instruction and Assessment

- Keep in mind that successfully responding to this probe requires that students understand that the Moon turns on its axis once a month and that one half of the Moon is illuminated by the Sun, no matter what phase it is in. Students might fail to understand one or both of these ideas, or they might have difficulty coordinating the two ideas while envisioning themselves on the surface of the Moon.
- If students struggle with this probe, conduct an activity so they can see what is happening on a physical model. Set up a bright light at the front of a darkened room. While the rest of the class watches from an imaginary viewpoint in space, give one student a ball to represent the Moon. Place a mark on the Moon ball to mark the location of the Moon base. The student's head represents the Earth, and the lightbulb represents the Sun. Have the class watch for the moment when the Moon base crosses the line between light and dark (corresponding to sunrise or sunset as viewed from the Moon). Meanwhile the student holding the Moon can call out phases while moving the ball in a circle, so students can see that the amount of the Moon lit by the Sun is independent of the phase as viewed from Earth.

★ Indicates a strong match between the ideas elicited by the probe and a national standard's learning goal.

- Do not assume that this probe is too easy for high school students. As indicated by the Trundle, Atwood, and Christopher study, even college students studying to become teachers have difficulty with these ideas. This probe will help you determine if it is necessary to spend time on this concept. If so, it is a good idea to use the modeling activity described above, because even high school students can benefit from using a hands-on model.
- Some of the most advanced students might also realize that the Moon must pass through Earth's shadow on occasion. It does not happen every month because the Moon's orbit is not in exactly the same plane as Earth's orbit around the Sun, but when it does happen (about twice a year) people living on the Moon would experience a solar eclipse lasting several hours.

References

American Association for the Advancement of Science (AAAS). 2009. Benchmarks for science literacy online. *www.project2061.org/publications/bsl/online*

Dove, J. 2002. Does the man in the Moon ever sleep? An analysis of student answers about simple astronomical events: A case study. *International Journal of Science Education* 24 (8): 823–834.

National Research Council (NRC). 1996. *National science education standards.* Washington, DC: National Academies Press.

Trundle, K., R. Atwood, and J. Christopher. 2002. Preservice elementary teachers' conceptions of Moon phases before and after instruction. *Journal of Research in Science Teaching* 39 (7): 633–658.

Does the Earth Go Through Phases?

Four students were working together on a science fiction story. Their story took place on the Moon. One of the students wrote a scene in which two of the characters were looking up at the "crescent Earth" just as it was setting just behind a mountain on the Moon. The students had different ideas about the scene. This is what they said:

Brett: "I don't think you would see phases of the Earth. The Earth always looks like a round blue ball from space. But I do think you would see the Earth set."

Margot: "I think we would see a 'crescent Earth.' It's just like being on Earth and looking up at the Moon. But the Earth would never 'set.'"

Scarlet: "I agree that we would see a 'crescent Earth.' The 'crescent Earth' would set, but it would set very slowly."

Hermione: "I don't think you would ever see a 'crescent Earth' or ever see the Earth rise or set."

Who do you think has the best idea about "Earth phases" and a rising and setting Earth? ______________________ Explain why you think that is the best idea.

____

____

____

Does the Earth Go Through Phases?

Teacher Notes

Purpose

The purpose of this assessment probe is to elicit students' ideas about "Earth phases" and "Earth sets." The probe is designed to find out if students understand the Earth-Sun-Moon system well enough to imagine how the Earth would appear as seen from the Moon.

Related Concepts

Moon: orbit, phase
Solar system objects: orbits

Explanation

Margot has the best answer: "I think we would see a 'Crescent Earth.' It's just like being on Earth and looking up at the Moon. But the Earth would never 'set.'"

The Earth will appear to go through phases when viewed from the Moon. As the Moon circles the Earth once a month, different parts of the lighted portion will come into view, so the Earth will appear to go through phases, just as the Moon appears to go through phases when seen from the Earth. However, the Earth would not appear to rise and set as seen from the Moon. People on the side of the Moon facing the Earth would always see the Earth in the same position, while people living on the far side would never see the Earth.

Administering the Probe

This is a challenging probe, best used with middle school or high school students in order to see if they are able to envision the Earth-Sun-Moon system from different points of view.

Related Ideas in *Benchmarks for Science Literacy* (AAAS 2009)

6–8 The Earth

- The Moon's orbit around the Earth once in about 28 days changes what part of the Moon is lighted by the Sun and how much

of that part can be seen from the Earth—the phases of the Moon.

Related Ideas in *National Science Education Standards* (NRC 1996)

5–8 Earth in the Solar System

- Most objects in the solar system are in regular and predictable motion. Those motions explain such phenomena as the day, the year, phases of the Moon, and eclipses.

Related Research

- Dai (1991) developed a multiple-choice questionnaire about the Moon, which included a question about phases of the Earth as viewed from the Moon. The questionnaire was validated by interviews, which provided richer insights into children's conceptions of the Earth-Sun-Moon system. The final version of the questionnaire was administered to 185 children in grades 5 and 6 in Taiwan. Only 34% of the students were found to hold a scientific understanding of the Earth-Sun-Moon system prior to instruction, with no differences due to age or gender. Following instruction involving field observations of the Moon, the use of physical models, and a play in which students acted out the movements of the Earth and Moon, students scored significantly higher than a control group. Nonetheless, many still held misconceptions about the Earth-Sun-Moon system, even after instruction. Dai concluded that the ability to envision the Earth-Sun-Moon system from space may be beyond the capabilities of most elementary school students.
- Barnett, Yamagata-Lynch, and Keating (2005) described a course for college undergraduates in which the students designed 3-D models of the solar system using virtual reality. In the process of designing their models, the students needed to imagine what would be seen from different viewpoints, such as looking back at the Earth from the Moon. When reporting on the results of the course, the researchers found that students who initially thought Earth would have phases as observed from the Moon were not able to articulate why they thought that was the case; but after designing their model to demonstrate what Earth phases would look like as seen from the Moon, they were able to explain quite clearly, as shown in the following example (Barnett, Yamagata-Lynch, and Keating 2005, p. 10):

Interviewer: Does the Earth have phases when viewed from the Moon?

Steve: Yes (using spheres to demonstrate his thinking). As the Moon moves around the Earth we go through full Earth, quarter Earth, new Earth, the same set of phases as the Moon has.

Jessica: Yes. It does. As the Moon goes around the Sun, sunlight hits the Earth. We definitely see phases of the Earth because our model shows it.

Suggestions for Instruction and Assessment

- After students have an opportunity to discuss the probe, have them act out the Earth-Sun-Moon system to test their initial ideas. Use a bright light in a darkened room and give each small group of students an Earth globe, or a ball representing the Earth. Suggest that the students take turns playing the part of the Moon.

That would involve sidestepping around the Earth, always keeping one face toward the Earth, and noting if (a) the Earth appears to go through phases (it will) and (b) the Earth appears to rise and set (it will not.) Afterward, have the students reconsider the probe. If they have changed their minds, ask the students when that change occurred—what helped them see the Earth-Sun-Moon system in a different way?

- This probe is challenging for students (and teachers!) of all ages, so it is a good idea to use it when students are ready, such as after a middle or high school course in astronomy. Use the activity described above to be sure students understand the Earth-Sun-Moon system before going on to more complex topics in astronomy.

References

American Association for the Advancement of Science (AAAS). 2009. Benchmarks for science literacy online. *www.project2061.org/publications/bsl/online*

Barnett, M., L. Yamagata-Lynch, and T. Keating. 2005. Using virtual reality computer models to support student understanding of astronomical concepts. *Journal of Computers in Mathematics and Science Teaching* 24 (4): 333–356.

Dai, M. 1991. Identification of misconceptions about the Moon held by fifth and sixth graders in Taiwan and an application of teaching strategies for conceptual change. Doctoral diss., University of Georgia.

National Research Council (NRC). 1996. *National science education standards.* Washington, DC: National Academies Press.

Is the Moon Falling?

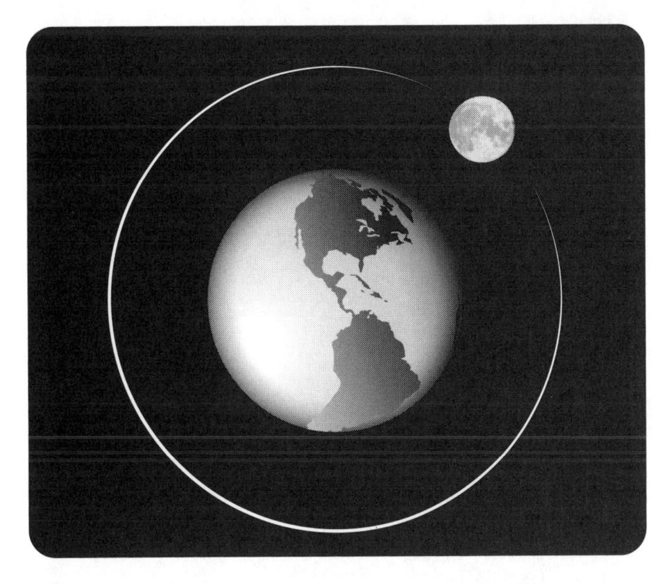

Maribel and Isaac are having an argument about the Moon. Here is what they said:

Maribel: "The Moon is falling toward the Earth because of gravity."

Isaac: "The Moon can't fall toward Earth because there is no gravity in space."

Circle whom you agree with the most: Maribel Isaac

Explain why you agree with one person and not the other. ______________________

____

____

____

____

____

____

____

Is the Moon Falling?

Teacher Notes

Purpose

The purpose of this assessment probe is to elicit students' ideas about gravity. The probe is designed to determine whether students understand the role of gravity in maintaining the Moon in its orbit.

Related Concepts

Gravity
Moon: orbit
Solar system objects: orbits

Explanation

Maribel has the best idea: "The Moon is falling toward Earth because of gravity." An object in orbit, such as the Moon, is actually falling due to the pull of gravity that acts between Earth and the object. Its sideways motion is just the right speed so that it neither falls to the surface nor flies off into space.

Isaac is wrong in that there *is* gravity in space. Although the force of gravity diminishes rapidly with distance, there is still sufficient gravitational force between the Earth and Moon to keep the Moon from flying off into space.

Administering the Probe

This probe is best used at the middle or high school level. However, keep in mind that understanding orbits is a difficult concept requiring that students have a good understanding of gravity. While students in middle school can develop a qualitative understanding of what keeps an object in orbit, details are usually addressed at the high school level.

Related Ideas in *Benchmarks for Science Literacy* (AAAS 2009)

3–5 Forces of Nature

- The Earth's gravity pulls any object on or near the Earth toward it without touching it.

6–8 Forces of Nature

★ The Sun's gravitational pull holds the Earth and other planets in their orbits, just as the planets' gravitational pull keeps their Moons in orbit around them.

9–12 Forces of Nature

- Gravitational force is an attraction between masses. The strength of the force is proportional to the masses and weakens rapidly with increasing distance between them.

Related Ideas in *National Science Education Standards* (NRC 1996)

5–8 Earth in the Solar System

★ Gravity is the force that keeps planets in orbit around the Sun and governs the rest of the motion in the solar system.

Related Research

- A review of 40 research studies about children's and adults' understanding of gravity (Kavanagh and Sneider 2006) noted that for students to make sense of how objects stay in orbit, they need to understand that gravitational forces extend beyond our own atmosphere, into space. Two different misconceptions lead students to believe that there is no gravity in space: (1) the idea that gravity needs air, so there can be no gravity in space where there is no air; and (2) the idea that Earth is a special place. The reviewed studies identified other related misconceptions: objects in orbit are weightless, so gravity does not affect them; the force of gravity diminishes rapidly with increasing altitude so it is absent above the atmosphere; an object must exert some force to maintain orbit; planets closer to the Sun or that spin faster have more gravity; and gravitational forces between objects are not equal and opposite. These misconceptions are surprisingly widespread, even among university students and teachers.
- Bar, Sneider, and Martimbeau (1997) conducted two 45-minute sessions for two classes of sixth graders on the role of gravity in keeping the Moon in orbit. Activities included observing the trajectories of balls rolling off tables, a series of slides showing how a cannonball could be fired fast enough to go into orbit, with extensions to the Space Shuttle and the Moon. The idea of gravity in space was addressed explicitly, noting that if gravity were suddenly "turned off," the Space Shuttle and Moon would fly off into space. Before the instruction 27% of students understood that there is gravity in space. After instruction the majority of students understood that there was gravity in space, at least "near planets," but 21% of the students still clung to the idea that there is no gravity in space.

Suggestions for Instruction and Assessment

- It is sufficient for students in upper elementary grades to learn that the Moon orbits Earth and that Earth and the other planets orbit the Sun. It is not necessary that they learn about the forces that keep planets in orbit.
- At the middle school level students are introduced to Newton's first and second

★ Indicates a strong match between the ideas elicited by the probe and a national standard's learning goal.

laws and learn about how forces, including gravity, cause objects to move. They may also learn that gravity keeps artificial satellites and the Moon in orbit, and they may develop a qualitative understanding of orbital motion.

- A common lesson on orbits is to have the students twirl a weight on the end of a string. The string represents the pull of gravity, keeping the object in orbit so it doesn't fly off into space. However, this demonstration is often not convincing to students who believe that there is no gravity in space, perhaps because of videos of astronauts shown floating "weightless" in their spacecraft. If students believe there is no gravity in space, consider using the probe "Talking About Gravity" in *Uncovering Student Ideas in Science, Vol. 1: 25 Formative Assessment Probes* (Keeley, Eberle, and Farrin 2005).
- Students can be introduced to orbits by observing the path taken by a ball that rolls off a table. The faster the ball is pushed, the farther from the table it will land. It is then possible to illustrate what would happen to a ball if thrown sideways faster and faster as it falls; it will eventually go into orbit. Isaac Newton illustrated this idea in a popular book, *The System of the World* (1728/1995), which described how gravity explains the orbit of the Moon. The image below is taken from his book and shows what would happen to an object launched horizontally at different speeds from the top of a huge mountain. It also shows that if an object is launched at a high enough rate of speed, it will go into orbit.

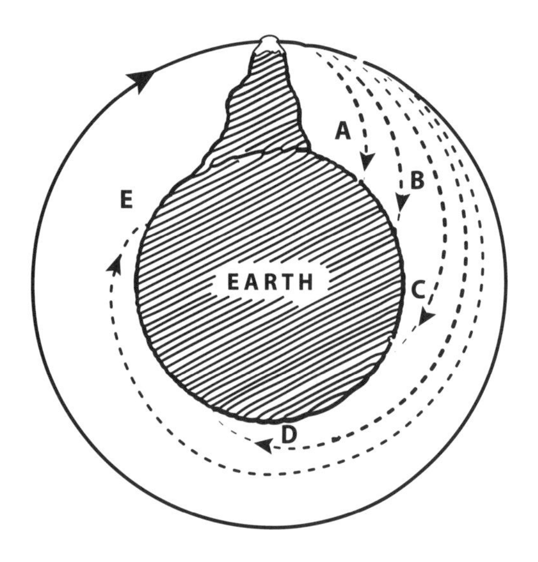

References

American Association for the Advancement of Science (AAAS). 2009. Benchmarks for science literacy online. *www.project2061.org/publications/bsl/online*

Bar, V., C. Sneider, and N. Martimbeau. 1997. What research says: Is there gravity in space? *Science and Children* 34 (7): 38–43.

Kavanagh, C., and C. Sneider. 2006. Learning about gravity II. Trajectories and orbits: A guide for teachers and curriculum developers. *Astronomy Education Review* 5 (2): 53–102. *http://aer.aas.org/resource/1/aerscz/v5/i2/p53_s1*

Keeley, P., F. Eberle, and L. Farrin. 2005. *Uncovering student ideas in science, vol. 1: 25 formative assessment probes.* Arlington, VA: NSTA Press.

National Research Council (NRC). 1996. *National science education standards.* Washington, DC: National Academies Press.

Newton, I. 1728. *The system of the world.* Amherst, NY: Prometheus Books, 1995.

Section 4

Dynamic Solar System

Concept Matrix: Dynamic Solar System
Probes 28–33

PROBES	28. What's Inside Our Solar System?	29. How Do Planets Orbit the Sun?	30. Is It a Planet or a Star?	31. Human Space Travel	32. Where Do You Find Gravity?	33. Gravity in Other Planetary Systems
GRADE-LEVEL USE →	3–8	3–12	3–12	6–12	6–12	6–12
RELATED CONCEPTS ↓						
apparent vs. actual size			X			
gravity		X		X	X	X
objects in the sky	X		X			
other planetary systems				X	X	
solar system objects: identity	X	X	X	X	X	
solar system objects: orbits		X			X	X
solar system objects: spin						X
space exploration				X	X	X
stars: locations				X		

Teaching and Learning Considerations

Textbooks, curriculum frameworks, and standards documents all include study of the solar system, usually at the upper elementary or middle school level, because knowledge of the solar system provides students with the big picture of our planet among other large bodies and many smaller ones, held together by gravitational attraction. Although this picture may not seem as relevant to daily life as many other topics in the curriculum, the concept of the solar system is an essential element of our cultural heritage; besides, children are fascinated by the Moon, planets, and stars. So for many the solar system provides a welcoming doorway into science.

Although the first two probes in this section appear simple, research studies show that people of all ages have misconceptions about what the solar system contains and how the planets move. These misconceptions provide useful diagnostic information about where to begin a unit on the solar system.

The next two probes ask students how to distinguish between a planet and a star and to share their understanding of space exploration. These probes will help you determine if students understand the solar system as a fanciful story that belongs in textbooks or as a system of actual bodies that can be seen in the sky and can be visited by robots and people in spacecraft.

The last two probes in this section concern gravity—the "glue" that holds the solar system together. These probes are designed to provide diagnostic information, as well as a springboard for discussion, to introduce the important connection between physical science and astronomy. As they discuss their ideas in response to these probes, your students may begin to see how the physical law that explains why a pencil falls on Earth is the same physical law that explains how a group of planets stay in orbit around a distant star.

Related Curriculum Topic Study Guides*

Gravity in Space
Motion of Planets, Moons, and Stars
Solar System
Space Technology and Exploration

*These guides are found in Keeley, P. 2005. *Science Curriculum Topic Study: Bridging the Gap Between Standards and Practice.* Thousand Oaks, CA: Corwin Press and Arlington, VA: NSTA Press. Each Curriculum Topic Study Guide provides a process to help the reader (1) identify adult content knowledge, (2) consider instructional implications, (3) identify concepts and specific ideas, (4) examine research on learning, (5) examine coherency and articulation, and (6) clarify state standards and district curriculum.

Related NSTA and Other Resources

NSTA Press Books

American Association for the Advancement of Science (AAAS). 2001. *Atlas of science literacy.* Vol. 1. (See "Gravity" map, pp. 42–43, and "Solar System" map, pp. 43–44.) Washington, DC: AAAS.

Gilbert, S. 2011. *Models-based science teaching.* Arlington, VA: NSTA Press.

Holt, G., and N. West. 2011. *Project Earth science: Astronomy.* 2nd ed. Arlington, VA: NSTA Press.

Keeley, P., F. Eberle, and L. Farrin. 2005. *Uncovering student ideas in science, vol. 1: 25 formative assessment probes.* (See "Talking About Gravity," pp. 97–102.) Arlington, VA: NSTA Press.

Keeley, P., and R. Harrington. 2010. *Uncovering student ideas in physical science, vol. 1: 45 new force and motion assessment probes.* Arlington, VA: NSTA Press.

NSTA Journal Articles

Brunsell, E., and J. Marcks. 2007. Teaching for conceptual change in space science. *Science Scope* 30 (9): 20–23.

Larson, B. 2009. Science shorts: Astronomies of scale. *Science and Children.* 47 (2): 54–56.

Lee, M., and D. Hanuscin. 2008. Perspective: A (mis)understanding of astronomical proportions. *Science and Children* 46 (1): 60–61.

Malonne, D., L. Landis, and A. Landis. 2009. Tried and true: Solar system in the hallway. *Science Scope* 32 (8): 46–49.

Riddle, B. 2002. Scope on the skies: Orbits and distances. *Science Scope* 26 (2): 8–9.

Riddle, B. 2008. Scope on the skies: Tracking planets around the Sun. *Science Scope* 31 (7): 84–86.

Riddle, B. 2010. Scope on the skies: The law of location. *Science Scope* 34 (1): 76–78.

Schuster, D. 2008. Take a planet walk. *Science and Children* 46 (1): 42–48.

NSTA Learning Center Resources

NSTA SciGuides

http://learningcenter.nsta.org/products/sciguides.aspx

Gravity and Orbits
The Solar System

NSTA SciPacks

http://learningcenter.nsta.org/products/scipacks.aspx

The Solar System

NSTA Science Objects

http://learningcenter.nsta.org/products/science_objects.aspx

Solar System, Asteroids, Comets, and Meteorites
Solar System: A Look at the Planets

Other Resources

Fraknoi, A., ed. 2011. *Universe at your fingertips 2.0.* San Francisco: Astronomical Society of the Pacific. Available at *www.astrosociety.org/uayf/index.html*

GEMS Space Science Sequence. Available from Carolina Biological Supply Company at *www.carolinacurriculum.com/GEMS*

What's Inside Our Solar System?

Many things are found inside our solar system. Put an X next to the things you think are found within our solar system.

_____**A** the Sun

_____**B** clouds

_____**C** galaxies

_____**D** the North Star

_____**E** Earth

_____**F** Earth's Moon

_____**G** planets

_____**H** Atlantic Ocean

_____**I** moons around other planets

_____**J** people

_____**K** comets

_____**L** small chunks of rock (meteoroids)

_____**M** large chunks of rock (asteroids)

_____**N** constellations

_____**O** United States

_____**P** airplanes

Explain your thinking. How did you decide which objects are inside our solar system?

____

____

What's Inside Our Solar System?

Teacher Notes

Purpose

The purpose of this assessment probe is to elicit students' ideas about the solar system. The probe is designed to find out if students can distinguish between objects in the solar system and objects outside of the solar system and recognize that things found on Earth are also found in the solar system.

Related Concepts

Objects in the sky
Solar system objects: identity

Explanation

The best answers are A, B, E, F, G, H, I, J, K, L, M, O, and P: the Sun, clouds, Earth, Earth's Moon, planets, the Atlantic Ocean, moons around other planets, people, comets, small chunks of rock (meteoroids), large chunks of rock (asteroids), the United States, and airplanes. These objects are all found within our solar system. Some, such as clouds, the Atlantic Ocean, the United States, and airplanes (answers B, H, O, and P) are found on Earth, which is within the solar system. The other objects listed on the probe—galaxies, the North Star, and constellations (answers C, D, and N)—are found outside of our solar system.

Administering the Probe

The probe is primarily designed for students in grades 3–5 who are just learning about the solar system. However, it is also appropriate for students in grades 6–8 who are studying astronomy, because researchers have found that many middle school children hold misconceptions such as the belief that other stars besides the Sun are present in the solar system or the belief that the Earth is not included in the solar system. You may remove or replace any objects on the list that students may be unfamiliar with. For example, *constellations* can be changed to *groups of stars* if students are not familiar with the word; the probe should not be an assessment of vocabulary. For older

students you might consider adding meteors, black holes, planets around other stars, dwarf planets, nebulae, pulsars, space junk, satellites, and other objects with which they have some familiarity.

This probe can also be used as a card sort (Keeley 2008). Print the objects on cards and have students work in small groups to sort the cards in three groups: (1) things we think are found in our solar system, (2) things we think are not found in our solar system, and (3) things we are unsure about or can't agree on. Students must defend their reasons for each card placement. Circulate and visually observe how students sort the cards while listening to their reasoning.

Related Ideas in *Benchmarks for Science Literacy* (AAAS 2009)

3–5 The Universe

- The Earth is one of several planets that orbit the Sun, and the Moon orbits around the Earth.

6–8 The Universe

★ Nine planets of very different size, composition, and surface features move around the Sun in nearly circular orbits. *[Note: This benchmark was written before Pluto was reclassified.]* Some planets have a variety of moons and even flat rings of rock and ice particles orbiting around them. Some of these planets and moons show evidence of geologic activity. The Earth is orbited by one moon, many artificial satellites, and debris.

★ Many chunks of rock orbit the Sun. Those that meet the Earth glow and disintegrate from friction as they plunge through the atmosphere—and sometimes impact the ground. Other chunks of rocks mixed with ice have long, off-center orbits that carry them close to the Sun, where the Sun's radiation (of light and particles) boils off frozen material from their surfaces and pushes it into a long, illuminated tail.

Related Ideas in *National Science Education Standards* (NRC 1996)

5–8 Earth in the Solar System

- The Earth is the third planet from the Sun in a system that includes the Moon, the Sun, eight other planets and their moons, and smaller objects such as asteroids and comets. *[Note: This standard was written before Pluto was reclassified.]* The Sun, an average star, is the central and largest body in the solar system.

Related Research

- Colombo, Aroca, and Silva (2010) administered a questionnaire and interviewed 137 students, ages 10–11, during a visit to a college observatory in Brazil. Although the students learned some information about the solar system during the visit, they were not able to develop a comprehensive view of the solar system. For example, 87% of the students responded that the Sun is a star, but only 18% thought the Sun was the only star in the solar system. Fewer than half of the students (45%) answered "yes" to the question, "Are you part of the solar system?"
- Sharp and Kuerbis (2005) compared two groups of 9- to 11-year-old children in England. One group received 20 hours of instruction on the solar system spread over 10 weeks, while the control group studied other topics. Each group had 31 children,

★ Indicates a strong match between the ideas elicited by the probe and a national standard's learning goal.

and there were no significant demographic differences between them. Each child was interviewed before and after the 10-week period, and children in the experimental group were interviewed again three months later. When asked what the solar system contained, prior to instruction all of the children named the Sun and planets, and a few also named moons, comets, meteorites, and asteroids. Only 3 of the 62 children interviewed could name all nine planets (the correct number at the time of the study). After instruction almost all of the children in the experimental group named moons, asteroids, meteorites, and comets and were able to describe these bodies, and all of the children knew there were nine planets. Children in the control group did not increase their knowledge of the solar system.

- Dussault (1999) reported on a survey of 257 visitors to the Smithsonian National Air and Space Museum in Washington, D.C., who were asked to name things found in the solar system. As expected, 82% named planets; but, surprisingly, 41% named stars and 18% named galaxies. Just 5% of the visitors named the Earth as a component of the solar system.
- Sadler (1992) developed a written test to measure high school students' understanding of astronomy concepts. The test was administered to 1,414 students in grades 8–12 who were just starting an Earth science or astronomy course. One of the questions was as follows (percentage of students who chose each answer is shown in parentheses):

  Which answer shows a pattern from closest object to the Earth to farthest from the Earth?
  A. Sun → Saturn → Moon (5%)
  B. Saturn → Moon → Sun (10%)
  C . Moon → Sun → Saturn (42%)
  D. Moon → Saturn → Sun (32%)
  E. Sun → Moon → Saturn (10%)
  Although 42% of the students gave the correct answer (C), a surprisingly large number of students thought that Saturn is closer to Earth than the Sun, even though it is between 3,500 and 4,500 times farther from Earth than the Sun. The researcher speculated that this misconception may arise from typical illustrations of the solar system in which the planets are all shown approximately the same distance from the Sun.

- Sadler et al. (2010) developed and validated a multiple-choice assessment item test bank for astronomy and space science in which the alternative answers for each question were based on research findings about common misconceptions. Results for a sample of 7,599 students and their 88 teachers are reported on the MOSART (Misconceptions-Oriented StandardsBased Assessment Resources for Teachers) website: *www.cfa.harvard.edu/smgphp/mosart/about mosart_2.html.* Items are listed for grades K–4, 5–8, and 9–12. Each question has five answer choices, so the odds of selecting the correct answer by chance are 20%. One item for grades 5–8 was as follows:

  Our solar system contains:
  A. One average star
  B. Several stars spread across space
  C. One older, dimmer star and one younger, brighter star
  D. Three stars
  E. No stars
  The correct response, A, was given by 32%. According to the researchers: "Most students selected B (50%). It may be that students have not yet conceptualized the difference between a star with its planets and a galaxy, which includes many stars."

Suggestions for Instruction and Assessment

- Students at all ages should begin studies leading to an understanding of the solar system in a way that is appropriate for their developmental level:
    - *At the primary level* the focus is on what students observe in the sky. Students should be trying to figure out what is in front of what, what changes in unpredictable ways (e.g., clouds), and what patterns are repeatable (e.g., rising and setting of the Sun).
    - *At the upper elementary level* students can begin to learn about solar system objects, beginning with the Sun and the Moon. Students should make a Moon calendar by recording their observations of the changing phases of the Moon. Students can also observe shadows and infer the regular apparent motion of the Sun across the sky. After becoming familiar with the Earth-Sun-Moon system, they can move to other objects, including planets. It is important for students to identify the easily visible planets in the sky (e.g., Jupiter, Mars, Saturn, Venus), learn to distinguish them from stars, and connect these actual planetary observations with the images of the planets taken with powerful telescopes and space probes.
    - *Upper elementary or middle school* is a good time for students to create a model of the solar system on the school grounds, using the same scale for both size and distance. Several modeling activities (e.g., "Peppercorn Earth") are available on the internet or as part of the rich packet of astronomy activities from the Astronomical Society of the Pacific (Fraknoi 2011; see "Activity D4. The Earth as a Peppercorn").
    - *At the high school level* students should have opportunities to study solar system objects in some detail, making systematic observations in the sky as well as accessing images and other resources via the internet. Understanding how the solar system functions as a gravitationally bound system can come together in the context of a physics class, when students study Newton's theory of universal gravitation.
- As students develop a conceptual understanding of the solar system and all that it contains, it is important to emphasize that Earth and everything on it—including people—are also in the solar system.
- Middle school is the time to sort out the various objects in the solar system and how they move in relation to each other. Students should have opportunities to view objects such as lunar craters, moons of Jupiter, and the rings of Saturn with a telescope or binoculars and learn about other celestial objects, such as comets and meteors, through videos that show how they are observed from Earth or via space probes.
- Be aware that some students may need to develop a conceptual understanding of solar system objects before encountering the terms. See, for example, the wording of the related 6–8 Benchmarks from the *Benchmarks for Science Literacy,* where the idea of asteroids and comets is developed first, conceptually, before introducing the terminology. You might consider doing the same if you add other objects to the list such as meteoroids.
- At all grade levels the focus should be on observing and how we know what we know about the solar system, rather than providing facts about solar system objects without relating how those facts are known.

References

American Association for the Advancement of Science (AAAS). 2009. Benchmarks for science literacy online. *www.project2061.org/publications/bsl/online*

Colombo, P., Jr., S. Aroca, and C. Silva. 2010. Daytime school guided visits to an astronomical observatory in Brazil. *Astronomy Education Review* 9 (1). *http://aer.aas.org/resource/1/aerscz/v9/i1/p010113_s1*

Dussault, M. 1999. How do visitors understand the universe? Studies yield information on planning exhibitions and programs. *ASTC Newsletter* (May/June): 9–11.

Fraknoi, A., ed. 2011. *Universe at your fingertips 2.0.* San Francisco: Astronomical Society of the Pacific. Available at *www.astrosociety.org/uayfl/index.html*

Keeley, P. 2008. *Science formative assessment: 75 practical strategies for linking assessment, instruction, and learning.* Thousand Oaks, CA: Corwin Press and Arlington, VA: NSTA Press.

National Research Council (NRC). 1996. *National science education standards.* Washington, DC: National Academies Press.

Sadler, P. M. 1992. The initial knowledge state of high school astronomy students. Doctoral diss., Graduate School of Education, Harvard University.

Sadler, P. M., H. Coyle, J. L. Miller, N. Cook-Smith, M. Dussault, and R. R. Gould. 2010. The Astronomy and Space Science Concept Inventory: Development and validation of assessment instruments aligned with the K–12 National Science Standards. *Astronomy Education Review* 8 (1). *http://aer.aas.org/resource/1/aerscz/v8/i1/p010111_s1.* Questions and findings for each question are available at the MOSART website: *www.cfa.harvard.edu/smgphp/mosart*

Sharp, J. G., and P. Kuerbis. 2005. Children's ideas about the solar system and the chaos in learning science. *Science Education* 90 (1): 124–147.

How Do Planets Orbit the Sun?

A teacher asked her students to list the order of the six planets that are closest to the Sun. All of the students were able to do this by listing:

Closest to Sun—**Mercury, Venus, Earth, Mars, Jupiter, and Saturn**—Farthest From Sun

Then the teacher surprised her students by asking them to draw the orbits of the planets, showing how they orbit around the Sun. Different students drew the orbits in different ways.

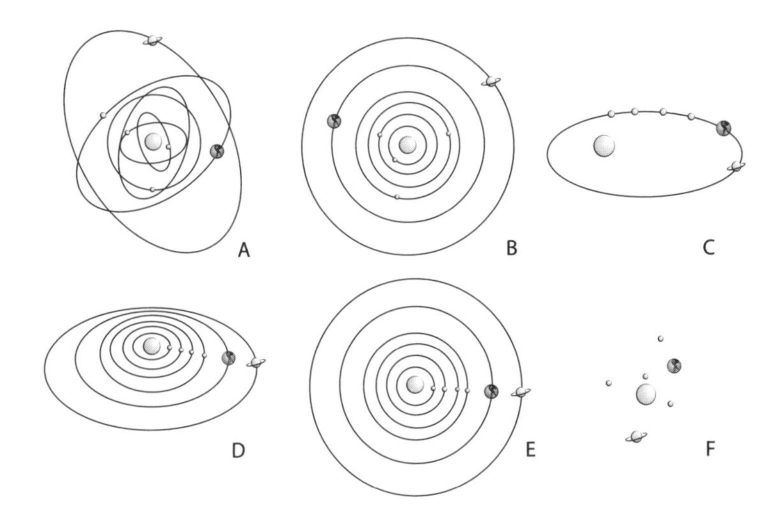

Which drawing do you think is most accurate? ________

Explain why you chose that drawing and not the others.

____

____

____

____

____

How Do Planets Orbit the Sun?

Teacher Notes

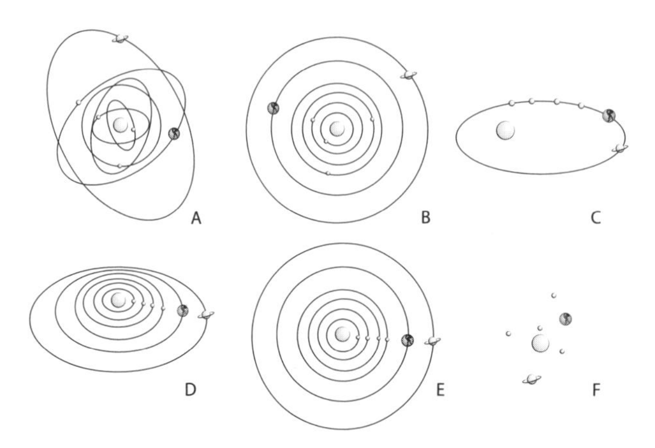

Purpose

The purpose of this assessment probe is to elicit students' ideas about representations of the solar system. The probe is designed to reveal students' ideas about the overall shape of the solar system and the planets within it.

Related Concepts

Gravity
Solar system objects: identity, orbits

Explanation

The best answer is B. This drawing illustrates that the major planetary orbits are nearly circular and concentric, and that the planets may be in any given position in their orbital paths. Also, drawing B shows the planets circling the Sun in nearly the same plane (called the ecliptic plane). Common misconceptions are: drawing A, which shows each planetary orbit in a different plane; drawing C, which shows the planets following each other in a single orbit; drawing D, which shows the planets in highly elliptical orbits and lined up on one side of the Sun; drawing E, which shows the correct shape of the orbits but has the planets lined up on one side of the Sun; and drawing F, which shows the planets arranged randomly around the Sun, including two gas giant planets (Jupiter and Saturn) closer to the Sun than some of the small rocky planets (Mercury, Venus, Earth, and Mars). Many children may prefer drawing D, since that is the way the solar system is often depicted in posters and textbook illustrations. Although all of the planets except Mercury do follow elliptical orbits, they are only slightly elliptical and appear more circular in shape, and the planets very rarely line up on one side of the Sun. Most artistic renderings of the solar system exaggerate the elliptical orbits to give the drawing depth, and show all the planets on one side of the Sun for convenience.

Administering the Probe

This probe is best used with students in grades 3–5 who are just learning about the solar system. However, it is also appropriate for students in grades 6–8 who are studying astronomy because researchers have found that many middle school children hold common misconceptions about planetary orbits. The probe can also be used with high school students as an elicitation prior to learning about Kepler's laws.

Related Ideas in *Benchmarks for Science Literacy* (AAAS 2009)

6–8 The Universe

★ Nine planets of very different size, composition, and surface features move around the Sun in nearly circular orbits. *[Note: This benchmark was written before Pluto was reclassified.]*

6–8 Forces of Nature

- The Sun's gravitational pull holds the Earth and other planets in their orbits, just as the planets' gravitational pull keeps their Moons in orbit around them.

Related Ideas in *National Science Education Standards* (NRC 1996)

5–8 Earth in the Solar System

- The Earth is the third planet from the Sun in a system that includes the Moon, the Sun, eight other planets and their moons, and smaller objects such as asteroids and comets. *[Note: This standard was written before Pluto was reclassified.]* The Sun, an average star, is the central and largest body in the solar system.

Related Research

- Sharp (1996) reported on interviews with 42 children, ages 10–11, who had learned about the solar system through England's national curriculum. The interviews revealed a variety of ideas about the structure of the solar system. Misconceptions included the planets being arranged in a single orbit, following each other around the Sun, following a spiral curve, or simply in a random arrangement.
- Sharp and Kuerbis (2005) compared the astronomical knowledge of two groups of students ages 9–11 in England by interviewing them before and after instruction. One group of 31 students participated in 10 two-hour astronomy lessons, while the other group of 31 studied other science topics. All students were interviewed before and after instruction. Before instruction nearly all students had little understanding of the solar system and a great majority expressed the common misconceptions about the planets reported by Sharp (1996). Only two of the children could correctly indicate the Sun at the center of the solar system, with planets arranged in concentric orbits in a plane around the Sun. There were few changes among those who did not participate in the astronomy lessons. However, those who did participate significantly improved their understanding. Twenty-two of the children held a heliocentric understanding of the solar system after instruction, and the great majority of those were aware that the planets orbited the Sun in the same direction, in roughly the same plane, and at different times. Furthermore, they understood that the planets remained in orbit because the Sun's gravity pulled them toward it.
- Neil Comins, in his book *Heavenly Errors: Misconceptions About the Real Nature of the Universe* (2003) and companion

★ Indicates a strong match between the ideas elicited by the probe and a national standard's learning goal.

website (*www.physics.umaine.edu/ncomins*) reported that some college students think that the Sun and all of the planets are arranged in a line. This idea may come from images that just show one half of the solar system, not drawn to scale, in which the planets are all lined up so they can be seen in one view. Other students think that planetary orbits are not in the same plane, but rather are randomly oriented.

- Sadler et al. (2010) developed and validated a multiple-choice assessment item test bank for astronomy and space science in which the alternative answers for each question were based on research findings about common misconceptions. Results were reported for a sample of 7,599 students and their 88 teachers on the MOSART (Misconceptions-Oriented Standards-Based Assessment Resources for Teachers) website: *www.cfa.harvard.edu/smgphp/mosart/aboutmosart_2.html.* Items are listed for grades K–4, 5–8, and 9–12. Since each question has five answer choices, the odds of selecting the correct answer by chance are 20%. One item from the grades 5–8 test bank was as follows:

  As Earth and Mars move they:

  A. Exchange positions with one another.
  B. Both get farther from the Sun than Jupiter.
  C. Move randomly through the solar system.
  D. Travel around the Sun with the Earth always closer.
  E. This isn't a good question because planets don't move.

  The correct response, D, was given by 54%. According to the researchers: "Students clearly believe that the orbits of Earth and Mars do not intersect. No incorrect response received significantly more student responses than the others (ranging between 9% and 13%)."

Suggestions for Instruction and Assessment

- Instruction in planetary orbits is best begun in middle school, when students are able to conceptualize Earth in space, orbiting the Sun along with the other major planets, and to learn about how gravity keeps all of the planets in nearly circular orbits.
- Point out to students that it is not a coincidence that all of the major planets and nearly all of their moons revolve around the Sun in the same direction and the same plane. The widely accepted explanation for this observation is that all of the planets were formed from a single huge rotating disc of gas and dust at about the time the Sun formed. The gravitational attraction between the various bits of material in the disc tended to accumulate into larger masses, which eventually became the major planets and their many moons. This process is nicely illustrated by a number of online videos, such as the NOVA program "Origin of the Solar System" (*http://video.pbs.org/video/1790621534/*).
- The challenge for high school students is to understand the balance of forces that result in stable orbits. The Teacher Notes for Probe 27, "Is the Moon Falling?" suggest an activity for introducing students to how a rocket can be shot into orbit and how gravity keeps the Moon in its orbit. This activity can be extended to illustrate how planets keep "falling" around the Sun. That is, they are attracted to the Sun, but since they are moving at a high rate of speed at right angles to the Sun's pull, they maintain a stable orbit.
- The balance of forces that result in stable orbits can be illustrated by tying a rub-

ber ball or tennis ball onto one end of a string, passing the string through a small tube, and hanging a weight on the opposite end of the string. By holding the tube it is possible to swing the ball in a circle so that it makes a circle around your hand. If it is swung faster, the orbit will be larger and the weight will be drawn up; if it is swung slower, the weight will pull the ball in closer to your hand. In this model the ball represents a planet, and the string and weight represent the pull of gravity.

References

American Association for the Advancement of Science (AAAS). 2009. Benchmarks for science literacy online. *www.project2061.org/publications/bsl/online*

Comins, N. F. 2003. *Heavenly errors: Misconceptions about the real nature of the universe.* New York: Columbia University Press.

National Research Council (NRC). 1996. *National science education standards.* Washington, DC: National Academies Press.

Sadler, P. M., H. Coyle, J. L. Miller, N. Cook-Smith, M. Dussault, and R. R. Gould. 2010. The Astronomy and Space Science Concept Inventory: Development and validation of assessment instruments aligned with the K–12 National Science Standards. *Astronomy Education Review* 8 (1). *http://aer.aas.org/resource/1/aerscz/v8/i1/p010111_s1.* Questions and findings for each question are available at the MOSART website: *www.cfa.harvard.edu/smg-php/mosart*

Sharp, J. 1996. Children's astronomical beliefs: A preliminary study of year 6 children in South-west England. *International Journal of Science Education* 18 (6): 685–712.

Sharp, J. G., and P. Kuerbis. 2005. Children's ideas about the solar system and the chaos in learning science. *Science Education* 90 (1): 124–147.

Is It a Planet or a Star?

One evening, just as the stars were starting to come out, Alphonse saw a bright object in the sky. It was there the next night and the next. Over the next few weeks when he watched for it at sunset he noticed that it appeared lower and lower toward the horizon. Eventually he didn't see it any more. When he told his friends about it, this is what they said:

Joseph: "You must have seen a planet."

Hector: "You must have seen a nearby star."

Hermione: "It was probably an airplane or satellite."

Which friend do you think has the best idea? ____________ Explain why you agree with that friend.

____

____

____

____

Is It a Planet or a Star?

Teacher Notes

Purpose

The purpose of this assessment probe is to elicit students' ideas about visible objects in the night sky. The probe is designed to reveal whether students know how to spot a planet and whether they can distinguish it from a star.

Related Concepts

Apparent vs. actual size
Objects in the sky
Solar system objects: identity

Explanation

Joseph has the best idea: "You must have seen a planet." Stars maintain their positions with respect to each other night after night. Planets, however, slowly change their positions with respect to the background stars. In fact, that's how they got their name—from the Greek *planētēs*, meaning "wanderer." Another difference is that stars appear to "twinkle" whereas planets do not.

Administering the Probe

The probe can be used with students in grades 3–12, although it is primarily designed for students in grades 3–5, when they first begin to learn about the planets. The probe reinforces the idea that we can see other planets from Earth and distinguish them from the stars. After students have discussed the probe and understand that Joseph is correct, ask them what Earth would look like as observed from another planet, such as Mars or Jupiter. (Earth would also appear to be a "star" that wandered slowly among stars that did not change their relative positions.)

Related Ideas in *Benchmarks for Science Literacy* (AAAS 2009)

3–5 The Universe

- The patterns of stars in the sky stay the same, although they appear to move across the sky nightly, and different stars can be seen in different seasons.
- ★ Planets change their positions against the background of stars.

Related Ideas in *National Science Education Standards* (NRC 1996)

K–4 Objects in the Sky

- The Sun, Moon, stars, clouds, birds, and airplanes all have properties, locations, and movements that can be observed and described.

K–4 Changes in Earth and Sky

- Objects in the sky have patterns of movement.

5–8 Earth in the Solar System

- Most objects in the solar system are in regular and predictable motion.

Related Research

- Sadler (1992) developed a written test to measure high school students' understanding of astronomy concepts. The test was administered to 1,414 students in grades 8–12 who were just starting an Earth science or astronomy course. One of the questions was as follows (percentage of students who chose each answer is shown in parentheses):

  Objects that can be seen with the unaided eye and appear to move against the background of stars during one month are always:

  A. farther away from us than the stars. (13%)
  B. within the solar system (39%)
  C. within the Earth's atmosphere. (24%)
  D. at the edge of the visible universe. (15%)
  E. a part of a binary star system. (7%)

  The correct answer is B, since the object described must be a planet within the solar system. Although that was the most common answer, the majority of children did not know how to distinguish a planet from its slow movement against the background stars.

- Sharp (1996) reported on interviews with 42 children, ages 10–11, who had learned about the solar system through England's national curriculum. The researcher reported that most children were confused about the difference between stars and planets. Although many of the children thought stars were like the Sun, others thought they were like planets or like the Moon. Also, very few children were aware that the stars appeared to move slowly during a single evening.
- Sadler et al. (2010) developed and validated a multiple-choice assessment item test bank for astronomy and space science in which the alternative answers for each question were based on research findings about common misconceptions. Results were reported for a sample of 7,599 students and their 88 teachers on the MOSART (Misconceptions-Oriented Standards-Based Assessment Resources for Teachers) website: *www.cfa.harvard.edu/smgphp/mosart/aboutmosart_2.html.* Items are listed for grades K–4, 5–8, and

★ Indicates a strong match between the ideas elicited by the probe and a national standard's learning goal.

9–12. Since each question has five answer choices, the odds of selecting the correct answer by chance are 20%. One item from the grade 5–8 test bank was as follows:

You go outside one night and see the pattern of the stars in the southern sky shown below.

Which of the views below shows how the stars would look 6 hours later?

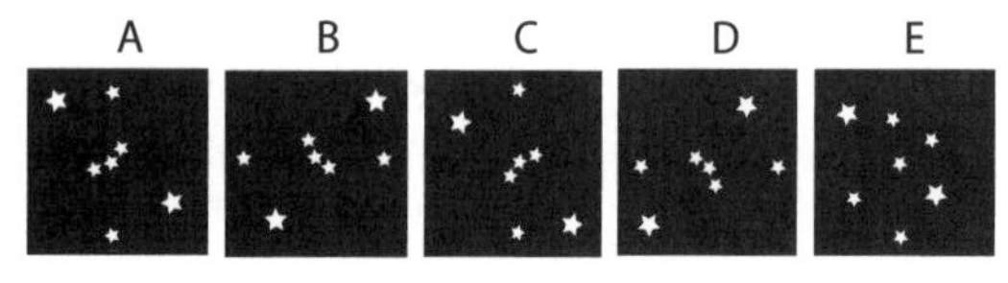

The correct response, B, was given by 20%. While the stars keep their relative positions with respect to each other, the entire constellation turns clockwise as the constellation moves from one side of the sky to the other. Students who observe the constellations for several hours during one evening would see this effect. According to the researchers: "The most frequent response to this question is A, which is visually identical to the pattern in the stem (40%). In addition, 19% chose C, a slight variation of the stem figure that could be seen as identical to the stem figure."

Suggestions for Instruction and Assessment

- Observations of planets in the night sky is best begun in upper elementary grades (4–5) when children can go outside after dark and observe the stars. If given as a homework assignment, it is best to give explicit instructions for how to find a given planet in the evening sky by drawing a sketch of the constellation where the planet will appear and telling the students which part of the sky to observe. There are many online sites that provide this information; see, for example, *www.skyandtelescope.com/observing/ataglance*. Be aware that some children living in urban areas with bright lights may not be able to see the stars or planets at night.
- If it is possible for the students to look through binoculars or a small telescope after they have located a planet in the sky, they will be able to see it as a small disc and, if the telescope is powerful enough, they may be able to see markings on its surface. In the case of Jupiter they will be able to see three or four of its moons. In the case of Saturn they may be able to see a hint of a ring. Such initial observations of the planets can be tremendously exciting for students of any age.
- By the time students reach middle school, they should be able to separate the movement of the stars in the nighttime sky from the movement of planets. Viewed from Earth the stars generally rise along the eastern horizon and set along the western horizon. But the stars maintain their positions relative to each other. The best way to observe that is to identify a few major constellations, like Orion, and to watch them early in the evening and again 3–4 hours later. During a single evening a planet among the stars appears to share the same motion; but viewed over several days and weeks, a planet does indeed "wander" among the stars. In order to identify planets, it is important for the students to first understand how the stars appear to move across the sky. Sadler et al.'s (2010) research question is relevant here, because it indicates that most students think the stars stay in the same position all night.

- An excellent project is to have your students carry out systematic naked-eye observations of a planet if they live in areas where the night sky is visible. Venus and Jupiter make the best targets because they are very bright and easy to spot. Students can make a sketch of the brightest nearby stars and then trace the movement of the planet against the stars every few days for a month or two. They can also trace the movement of the Moon against background stars. The Moon will be much easier to see and will appear to move more rapidly, since it is closest to Earth. Point out how the ancient Greek and Roman astronomers considered the Sun and Moon to be "planets" because they also gradually changed their positions against the background stars.

References

American Association for the Advancement of Science (AAAS). 2009. Benchmarks for science literacy online. *www.project2061.org/publications/bsl/online*

National Research Council (NRC). 1996. *National science education standards.* Washington, DC: National Academies Press.

Sadler, P. M. 1992. The initial knowledge state of high school astronomy students. Doctoral diss., Graduate School of Education, Harvard University.

Sadler, P. M., H. Coyle, J. L. Miller, N. Cook-Smith, M. Dussault, and R. R. Gould. 2010. The Astronomy and Space Science Concept Inventory: Development and validation of assessment instruments aligned with the K–12 National Science Standards. *Astronomy Education Review* 8 (1). *http://aer.aas.org/resource/1/aerscz/v8/i1/p010111_s1.* Questions and findings for each question are available at the MOSART website: *www.cfa.harvard.edu/smg-php/mosart*

Sharp, J. 1996. Children's astronomical beliefs: A preliminary study of year 6 children in South-west England. *International Journal of Science Education* 18 (6): 685–712.

Human Space Travel

An American astronaut, Neil Armstrong, was the first person to walk on the Moon. He made his historic Moon walk in 1969. Several decades have passed since the Apollo astronauts walked on the Moon. What do you think is the farthest distance humans have traveled in space since the year 2000? Circle the answer you think best describes the farthest distance astronauts have traveled recently.

A into the upper part of Earth's atmosphere

B about a quarter of the way to the Moon

C about halfway to the Moon

D to the Moon

E slightly past the Moon

F halfway to Mars

G to Mars

H slightly past Mars

I almost halfway through our Solar System

J to a nearby star

K to another planetary system

L to another galaxy

Explain your thinking. How did you decide how far recent astronauts have traveled?

____

____

____

Human Space Travel

Teacher Notes

Purpose

The purpose of this assessment probe is to elicit students' ideas about space travel. The probe is designed to reveal how far students think humans have traveled in space since the year 2000. The probe also gives a glimpse into students' understanding of relative distances in space.

Related Concepts

Gravity
Other planetary systems
Solar system objects: identity
Space exploration
Stars: locations

Explanation

The best answer is A: into the upper part of Earth's atmosphere. Human space travel during the last few decades has been through the Space Shuttle program or through Russian and other international space programs that send astronauts to the International Space Station (ISS). The Space Shuttle and the ISS both orbit the Earth in the part of our atmosphere called the thermosphere. This part of the atmosphere extends 220–500 miles (350–800 km) above Earth's surface. (There is no clearly defined altitude at which the atmosphere simply ends. The density of air simply gets thinner and thinner with increasing altitude. Although the air is exceedingly thin at that altitude, it is still considered the upper atmosphere.) Astronauts have not traveled very far in the last few decades compared with the distance traveled during the Apollo rocket era.

Students may give incorrect answers to this question for very different reasons. The first is that they simply have not learned about the lunar landings of the 20th century. From their perspective Armstrong's walk on the Moon is ancient history, and if they did not study the history of space exploration, they may not know how far humans have traveled in space or whether the famous Mars Rovers carried human crews. Second, research has shown that

many students are confused about the relative sizes of and distances to various celestial objects. So, for example, if they responded "to a nearby star," they may be thinking that stars are within the solar system. Third, many students may be unaware of the difficulty and immense cost of flying humans into space, and they may believe that space travel is as easy as illustrated in science fiction stories. Fourth, since students likely know that astronauts cannot breathe in space, they could reasonably conclude that astronauts have left the atmosphere.

Administering the Probe

The probe is best used with students in grades 6–12. Make sure students understand that the question is asking how far astronauts have traveled from the year 2000 until now. In other words what is the farthest people have gone during this time frame? Consider following up this probe with a discussion on limitations that affect how far humans can travel in space.

Related Ideas in *Benchmarks for Science Literacy* (AAAS 2009)

6–8 The Universe

- Nine planets of very different size, composition, and surface features move around the Sun in nearly circular orbits. *[Note: This benchmark was written before Pluto was reclassified.]* Some planets have a variety of moons and even flat rings of rock and ice particles orbiting around them. Some of these planets and moons show evidence of geologic activity. The Earth is orbited by one moon, many artificial satellites, and debris.

9–12 The Universe

- Increasingly sophisticated technology is used to learn about the universe. Visual, radio, and x-ray telescopes collect information from across the entire spectrum of electromagnetic waves; computers handle data and complicated computations to interpret them; space probes send back data and materials from the remote parts of the solar system; and accelerators give subatomic particles energies that simulate conditions in the stars and in the early history of the universe before stars formed.

6–8 Technology and Science

- Technology is essential to science for such purposes as access to outer space and other remote locations, sample collection and treatment, measurement, data collection and storage, computation, and communication of information.

9–12 Technology and Science

- One way science affects society is by stimulating and satisfying people's curiosity and enlarging or challenging their views of what the world is like.

Related Ideas in *National Science Education Standards* (NRC 1996)

5–8 Earth in the Solar System

- The Earth is the third planet from the Sun in a system that includes the Moon, the Sun, eight other planets and their moons, and smaller objects such as asteroids and comets. *[Note: This standard was written before Pluto was reclassified.]* The Sun, an average star, is the central and largest body in the solar system.

Related Research

- Sadler (1992) developed a written test to measure high school students' understanding of astronomy concepts, and gave it to

1,414 students in grades 8–12 who were just about to start a course in Earth science or astronomy. One question concerned the relative distances to the stars, Pluto, and the Space Shuttle in orbit. The correct answer (that the Space Shuttle is closer than the stars and Pluto) was given by 44% of the students; 26% responded that the stars are closer than Pluto, and 17% indicated that the Space Shuttle went out beyond the stars. According to the researcher: "The results of this question support the idea that many students believe that there are stars within the solar system, between the Earth and Pluto. Some students actually believe that the Space Shuttle goes out beyond the stars. I have talked to students who are mystified about the reason why our spaceships have not visited other solar systems, since they think our spacecraft can reach them. Many of these students can see no impediments to human colonization of the galaxy and view the possibility that visitors from other solar systems have come to the Earth as totally reasonable."

- Sadler et al. (2010) developed and validated a multiple-choice assessment item test bank for astronomy and space science in which the alternative answers for each question were based on research findings about common misconceptions. Results were reported for a sample of 7,599 students and their 88 teachers on the MOSART (Misconceptions-Oriented Standards-Based Assessment Resources for Teachers) website: *www.cfa.harvard.edu/smgphp/mosart/aboutmosart_2.html*. Items are listed for grades K–4, 5–8, and 9–12. Since each question has five answer choices, the odds of selecting the correct answer by chance are 20%. Two relevant items from the grades 5–8 test bank were as follows:

1. Of the locations below, the most distant place we have had a spacecraft fly by is:
   A. the Moon
   B. Mars
   C. Neptune
   D. the star Betelgeuse
   E. the Andromeda galaxy
   The correct answer, C, was given by 31%. According to the researchers: "The most popular response to this item was B, with 37% indicating that spacecraft that have flown by Mars. This may be a response to the publicity accorded the Mars Rovers. Only a very small minority thought that spacecraft had left the solar system (choices D and E, 10% and 8%)." *[Note: As this book goes to print, the* Voyager 1 *and* 2 *robotic spacecraft are both well beyond the orbit of the dwarf planet Pluto.]*

2. Which answer shows the most accurate pattern of the three objects in order from closest object to Earth to farthest from Earth?
   A. Space Shuttle in orbit → stars → Pluto
   B. Pluto → Space Shuttle in orbit → stars
   C. Stars → Space Shuttle in orbit → Pluto
   D. Stars → Pluto → Space Shuttle in orbit
   E. Space Shuttle in orbit → Pluto → stars
   The correct answer, E, was given by 43%. According to the researchers: "Although the correct answer was chosen most often, most students chose other sequences, suggesting that many students lack a coherent model of distances within the solar system."

- Dussault (1999) reported that surveys of visitors conducted to help plan science center exhibits found that few adult visitors were aware of the relative distances of various celestial objects, often placing stars and galaxies within the solar system and assuming that the Hubble Space Telescope is beyond the Moon, rather than in low Earth orbit.

Suggestions for Instruction and Assessment

- Although elementary students may enjoy talking about space travel, it is not likely they can develop a realistic understanding of where things are located in space, relative sizes and distances of celestial objects, or the vastly different requirements for vehicles that can support human versus robotic expeditions. It is therefore best to save this probe until middle school or high school.
- Middle school is the ideal time to introduce students to the real story of space travel. There are many instructional DVDs and even feature-length movies (such as *Apollo 13*) that realistically portray the history of human spaceflight. Such lessons are of even greater value if they are accompanied by instruction on the solar system so that the students can recognize what a relatively small step has been accomplished by space voyagers so far, yet how difficult and dangerous even those initial efforts were.
- A number of curriculum materials feature what has been learned by human exploration of space, such as the FOSS (Full Option Science System) Planetary Science module that engages students in thinking about the scientific payoff of the Apollo program: *http://lhsfoss.org/scope/folio/html/PlanetaryScience/1.html*. NASA has a number of websites that present the findings of space missions aimed at understanding human impacts on Earth systems; one example is the Earth Observatory at *http://Earthobservatory.nasa.gov*.
- Another approach is to focus on the many technological spin-offs of space exploration; see, for example: *www.spaceexplorationday.us/benefits/technology.html*.
- It is also important that space exploration not become just a history lesson. Students need to become aware of current efforts to explore space by astronauts on board the ISS and of the work of private companies and people in other countries to develop space vehicles.
- Whether this information is presented through a physics course, an astronomy course, or an integrated science course, high school students should graduate with a broad understanding of humankind's place within the solar system, the galaxy, and the universe as a whole. They should also be aware of human and robotic explorations of space, why they were undertaken, and what they have accomplished.
- This probe can be extended to ask students what is the farthest distance spacecraft (with or without humans on board) have traveled or where they think the Hubble Space Telescope is found.

References

American Association for the Advancement of Science (AAAS). 2009. Benchmarks for science literacy online. *www.project2061.org/publications/bsl/online*

Dussault, M. 1999. How do visitors understand the universe? Studies yield information on planning exhibitions and programs. *ASTC Newsletter* (May/June): 9–11.

National Research Council (NRC). 1996. *National science education standards.* Washington, DC: National Academies Press.

Sadler, P. M. 1992. The initial knowledge state of high school astronomy students. Doctoral diss., Graduate School of Education, Harvard University.

Sadler, P. M., H. Coyle, J. L. Miller, N. Cook-Smith, M. Dussault, and R. R. Gould. 2010. The Astronomy and Space Science Concept Inventory: Development and validation of assessment instruments aligned with the K–12 National Science Standards. *Astronomy Education Review* 8 (1). *http://aer.aas.org/resource/1/aerscz/v8/i1/p010111_s1.* Questions and findings for each question are available at the MOSART website: *www.cfa.harvard.edu/smgphp/mosart*

Where Do You Find Gravity?

Most people know that an apple falls to the ground because of Earth's gravity. Put an X next to all the other places where gravity exists.

_____ Earth's atmosphere

_____ just outside of Earth's atmosphere

_____ the Moon

_____ Mars

_____ Jupiter

_____ Pluto

_____ Sun

_____ distant stars

_____ galaxies

_____ far out in the distant universe

Explain your thinking. What rule or reasoning did you use to decide where gravity exists?

____

____

____

____

____

Where Do You Find Gravity?

Teacher Notes

Purpose

The purpose of this assessment probe is to elicit students' ideas about gravity. The probe is designed to see if students recognize that gravity exists everywhere in space, not just on Earth.

Related Concepts

Gravity
Other planetary systems
Solar system objects: identity, orbits
Space exploration

Explanation

Every place on the list should be checked. Gravity is a property of all objects with mass, so wherever there is mass—no matter how small—there will be gravity. Even a tiny electron exerts a gravitational pull. And although gravitational force diminishes rapidly as distance from a mass increases, gravitational force does not disappear completely, so that even in the vast empty spaces between galaxies there is some gravity.

Administering the Probe

This probe is best used with grades 6–12. The probe can be used as a card sort (Keeley 2008), with choices printed on cards. Students work in small groups to sort the cards into places where gravity exists, places where gravity does not exist, and places they are unsure about or do not agree on, justifying their reasons for each choice. As you circulate, you can easily see at a glance whether students recognize that gravity exists everywhere and listen in as they discuss their reasoning.

Related Ideas in *Benchmarks for Science Literacy* (AAAS 2009)

6–8 Forces of Nature

★ Every object exerts gravitational force on every other object. The force depends on how much mass the objects have and on how far apart they are. The force is hard to detect unless at least one of the objects has a lot of mass.

★ The Sun's gravitational pull holds the Earth and other planets in their orbits, just as the planets' gravitational pull keeps their moons in orbit around them.

Related Ideas in *National Science Education Standards* (NRC 1996)

9–12 Motions and Forces

★ Gravitation is a universal force that each mass exerts on any other mass.

Related Research

- Bar, Sneider, and Martimbeau (1997) conducted a learning study with two classes of sixth graders, including a total of 48 students. All students were given written pre- and posttests, and 10 of the students were interviewed before and after the intervention. Before instruction students were asked: "Does gravity act in space where there is no air?" Only 27% of the students thought that it did so. Instruction consisted of two class periods in which students observed the motion of balls rolling off of tables, forming a longer arc when pushed harder. A series of slides illustrated how the arc could be increased even further and eventually put an object into orbit where it would fall, but miss the Earth entirely. The instruction was successful with many but not all of the students. After instruction 48% of the students agreed that there was gravity in space, and an additional 15% answered that there was just a little gravity in space, or it existed just near planets.
- Ruggiero et al. (1985) interviewed 22 Italian middle school students, ages 12–13 years, about astronomy. Although none of the questions explicitly mentioned air, several students indicated that air is needed for gravity to function: "Weight exists only on the Earth, because there is air … and gets smaller higher up because air is rarefied" (Ruggiero et al., pp. 187–188). When asked to describe the motion of objects on the Moon, some students said that objects there floated because of the absence of gravity. Some students believed that the lack of air on the Moon meant that there was no gravity there. Others viewed gravity as something that is both Earth-centered and Earth-specific. The Moon was considered to be too far from the Earth to experience gravity.
- Noce, Torosantucci, and Vicentini (1988) administered a questionnaire to 88 Italian fifth graders about an astronaut who loses a tool on the Moon. Only 8 of the children said the tool would fall, and of those, only one said it would fall because it is attracted by gravity. The great majority of the students' answers indicated that gravity only applied on Earth. For example: "The force of gravity is our atmosphere." "It is a kind of air that pulls downwards." "[It is a] force of the universe that pushes people down."
- In 1993, Reynoso, Fierro, and Torres administered questionnaires and conducted interviews with 302 Mexican students from preschool through pre-university as well as 20 teachers. Respondents were asked about their beliefs regarding the motion of

★ Indicates a strong match between the ideas elicited by the probe and a national standard's learning goal.

objects falling on Earth and on the Moon. The researchers found that students largely maintained the belief that objects would fall on Earth and would float if released on the Moon. Older students tended to use more sophisticated language to explain their reasoning, whether their conceptual understanding was correct or not.

Suggestions for Instruction and Assessment

- This probe can be combined with "Talking About Gravity" (Keeley, Eberle, and Farrin 2005) and "Experiencing Gravity" (Keeley and Harrington 2010). Note that *Uncovering Student Ideas in Physical Science, Vol. 1: 45 New Force and Motion Assessment Probes* (Keeley and Harrington 2010) includes an entire section with gravity probes.
- The probe is a good way of bringing to the surface students' ideas about how gravity acts in space. Summarize the discussion by identifying three or four alternative ideas. It is best to leave those ideas as hypotheses to consider rather than pronouncing one as correct at this stage. Students will be much more interested in lessons on gravity if there is some mystery to uncover.
- By the end of fifth grade, students should have a fairly good image of a spherical Earth in space, with people living all over the Earth, held on the surface by gravity pulling toward the center of the Earth. That will provide a good foundation for learning at the middle school level about how gravity acts in space.
- One of the reasons that students have difficulty with the idea that there is gravity in space is that they have seen images of astronauts floating "weightless" in the Space Shuttle. This is an unfortunate use of the term, since *weight* is defined in most textbooks as the force of gravity on an object. The Space Shuttle is only about 200 miles above the surface of the Earth, so that is only a 5% reduction in gravitational force. A more accurate description is that the astronauts are "in free fall," which means that they are falling to Earth at the same rate as the Space Shuttle and everything in it. Students can be introduced to this idea as explained in the Teacher Notes for Probe 27, "Is the Moon Falling?" in this volume.
- High school physics students will learn to calculate the force of gravity between any two objects given their masses and the distance between their centers of gravity. However, being able to solve a calculation is not the same as having a conceptual understanding of how gravity acts between two objects. High school students are as likely as middle school students to have difficulty understanding the idea of free fall in orbit or the idea that the force between two masses acts equally on each mass (e.g., the force that you exert on the Earth is the same as the force that Earth exerts on you, as dictated by Newton's third law of motion).

References

American Association for the Advancement of Science (AAAS). 2009. Benchmarks for science literacy online. *www.project2061.org/publications/bsl/online*

Bar, V., C. Sneider, and N. Martimbeau. 1997. Is there gravity in space? *Science and Children* 34 (4): 38–43.

Keeley, P. 2008. *Science formative assessment: 75 practical strategies for linking assessment, instruction, and learning.* Thousand Oaks, CA: Corwin Press and Arlington, VA: NSTA Press.

Keeley, P., E. Eberle, and L. Farrin. 2005. *Uncovering student ideas in science, vol. 1: 25 formative assessment probes.* Arlington, VA: NSTA Press.

Keeley, P., and R. Harrington. 2010. *Uncovering student ideas in physical science, vol. 1: 45 new*

force and motion assessment probes. Arlington, VA: NSTA Press.

National Research Council (NRC). 1996. *National science education standards.* Washington, DC: National Academies Press.

Noce, G., G. Torosantucci, and M. Vicentini. 1988. The floating of objects on the Moon: Prediction from a theory of experimental facts? *International Journal of Science Education* 10 (1): 61–70.

Reynoso, E., E. Fierro, and G. Torres. 1993. The alternative frameworks presented by Mexican students and teachers concerning the free fall of bodies. *International Journal of Science Education* 15 (2): 127–138.

Ruggiero, S., A. Cartelli, F. Dupre, and M. Vincentini-Missoni. 1985. Weight, gravity, and air pressure: Mental representations by Italian middle school pupils. *European Journal of Science Education* 7 (12): 181–194.

Gravity in Other Planetary Systems

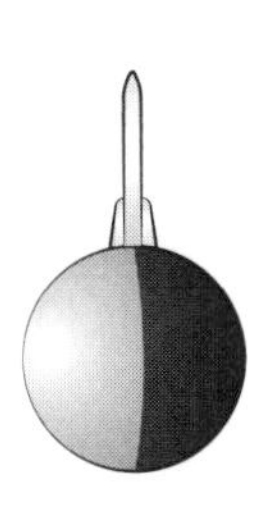

PLANET A
Fast Spin
Low Mass
Close to Sun

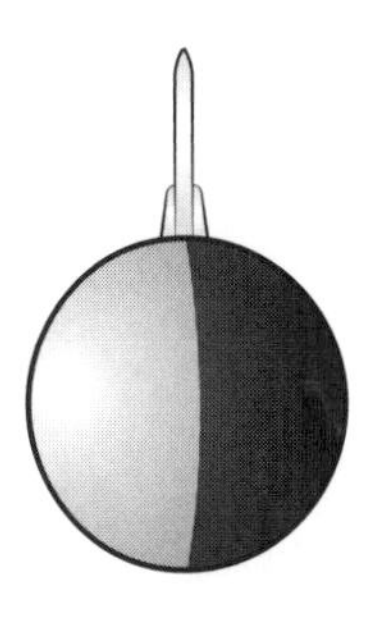

PLANET B
Slow Spin
High Mass
Far from Sun

(Rocket and planet are not to scale.)

Simon liked to dream about the day that people would travel to other worlds in other planetary systems. One day he got to thinking about what it might be like to take off in a rocket from planets that were different sizes and different distances from their sun. He drew a picture and showed it to some friends. He wanted to get their ideas about which rocky planet would be hardest for the rocket to take off from. Here is what his friends said:

Juanita: "No difference. It will be the same for both planets."

Rufus: "I think the rocket would have to work harder to get off planet A."

Jojo: "I think the rocket would have to work harder to get off planet B."

Whom do you agree with the most? ____________ Explain why you agree.

____

____

____

____

Gravity in Other Planetary Systems

Teacher Notes

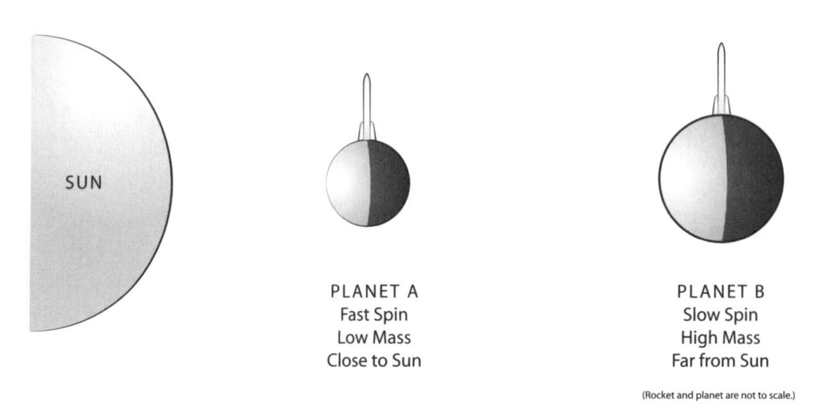

Purpose

The purpose of this assessment probe is to elicit students' ideas about gravity. The probe is designed to find out if students recognize the effect of mass on gravitational attraction, and that neither a planet's spin nor its distance from the system's central star will affect its gravitational attraction.

Related Concepts

Gravity
Solar system objects: orbits, spin
Space exploration

Explanation

Jojo has the best answer: "I think the rocket would have to work harder to get off planet B." Planet B has the greatest mass, so it exerts the most gravitational force, which affects the attraction of the rocket to the planet. Gravity is directly proportional to mass. If an object has twice as much mass, it has twice as much gravitational attraction to any other mass. Common misconceptions are that planets that spin faster on their axis or that are closer to their sun will have more gravity. Students with a sophisticated understanding of orbital dynamics may point out that the spin of a planet could increase or decrease the amount of thrust needed to get a rocket into orbit, depending on the location of the launch site and direction of the orbit, but the effect of planetary spin is much smaller than the effect due to the mass of the planet.

Administering the Probe

This probe is primarily designed for students in middle and high school. Make sure students know this is an imaginary situation. Additionally, they are being asked to address multiple variables at the same time for the purpose of getting at some of the common misconceptions about proximity to the Sun and spin being major factors affecting gravity. Listen carefully

to students' responses; they may have legitimate justification for alternative responses. For example, a knowledgeable student could note that a quickly spinning planet is easier to take off from because the spinning motion gives the object some of the speed necessary to leave the planet.

Related Ideas in *Benchmarks for Science Literacy* (AAAS 2009)

6–8 Forces of Nature

- ★ Every object exerts gravitational force on every other object. The force depends on how much mass the objects have and on how far apart they are. The force is hard to detect unless at least one of the objects has a lot of mass.
- The Sun's gravitational pull holds the Earth and other planets in their orbits, just as the planets' gravitational pull keeps their moons in orbit around them.

Related Ideas in *National Science Education Standards* (NRC 1996)

9–12 Motions and Forces

- ★ Gravitation is a universal force that each mass exerts on any other mass.

Related Research

- Treagust and Smith (1986) and Smith and Treagust (1988) surveyed Australian 10th graders after a series of lessons in astronomy. Twenty-four students were interviewed and 113 students were given a questionnaire to fill out. The students were enthusiastic about the astronomy lessons and had retained some knowledge, including the names of heavenly objects and our location in the universe. However, the students were unable to convey fundamental conceptual knowledge about how gravity works and how the orbit of a planet is affected by its location in the solar system. One of the questions showed a drawing of three planets: one with no rotation, one with slow rotation, and one with fast rotation. Students were asked, "Which planet will be easiest to 'take off' from?" Forty-seven percent of the students responded that the planet with no rotation would be easiest to take off from. Sixteen percent chose the planet with slow rotation, "since Planet B's rotations are not too fast and does not hold things down." In the same study, many of the students expressed the idea that planets closer to the Sun had stronger gravity. The researchers concluded, "Even some of the more able students expressed these misunderstandings and misconceptions" (Treagust and Smith 1986, p. 390).
- Piburn (1988) interviewed and administered written questionnaires about astronomy and logic to 40 Australian university students who were nonscience majors and found that many of the misconceptions expressed by high school students persisted at the university level. In response to a question about gravity, many of the students agreed with the statement "The gravity of a planet depends on its distance from the Sun." Responses to this question were closely correlated with performance on the logic test: 60% of the students who performed poorly on the logic test agreed with the statement, while only 27% of the students who performed well on the logic test agreed.

★Indicates a strong match between the ideas elicited by the probe and a national standard's learning goal.

Suggestions for Instruction and Assessment

- It is sufficient for upper elementary students to develop a simple understanding of gravity as a force that pulls everything toward Earth's center, so that people can live all over the Earth without falling off. Save the discussion of gravity in space until middle school.
- Use the probe to engage the students in discussion about what makes gravity stronger on some planets and weaker on other planets. Is it the closeness of the planet to the Sun? The rate of spin? Or just the mass of the planet?
- You can use one of the many websites that provide information on the gravitational attraction of various planets. For example, the Exploratorium at *www.exploratorium.edu/ronh/weight* allows students to enter their own weight and see how much they would weigh on different planets or the Moon, the Sun, or a star. They can probably infer from that experience that mass is the most important factor.
- High school physics students will learn to calculate the force of gravity between any two objects given their masses and the distance between their centers of gravity. However, it will also be important for the students to develop a conceptual understanding of gravity by discussing situations like that depicted in this probe. Other scenarios they might discuss include:
    - If Earth were to stop spinning would gravity increase, decrease, or stay the same? (It would stay the same. However, if the Earth stopped spinning, very sensitive scales would report a small difference in weight. This is not because gravity is actually less, but because of the centripetal effects of the Earth spinning.)
    - If Earth were to move closer to the Sun, would gravity increase, decrease, or stay the same? (It would stay the same.)
    - If a large body from space were to merge with the Earth, would gravity increase, decrease, or stay the same? (It would increase because Earth would be more massive.)

References

American Association for the Advancement of Science (AAAS). 2009. Benchmarks for science literacy online. *www.project2061.org/publications/bsl/online*

National Research Council (NRC). 1996. *National science education standards.* Washington, DC: National Academies Press.

Piburn, M. 1988. Misconceptions about gravity held by college students. Paper presented at the National Association for Research in Science Teaching, Lake of the Ozarks, Missouri. (ERIC Document Reproduction no. ED 292616)

Smith, C., and D. Treagust. 1988. Secondary students' understanding of gravity and the motion of planets. *School Science and Mathematics* 89 (5): 380–391.

Treagust, D., and C. Smith. 1986. Secondary students' understanding of the solar system: Implications for curricular revision. Paper presented at the annual conference of the International Group for the Advancement of Physics Teaching, Copenhagen, Denmark.

Section 5

Stars, Galaxies, and the Universe

Concept Matrix: Stars, Galaxies, and the Universe
Probes 34–45

PROBES	34. Shooting Star	35. Is the Sun a Star?	36. Where Are the Stars in Orion?	37. Which Is Bigger?	38. What Are Stars Made Of?	39. What Happens to Stars When They Die?	40. Do Stars Change?	41. Are We Made of Star Stuff?	42. Seeing Into the Past	43. What Is the Milky Way?	44. Expanding Universe	45. Is the Big Bang "Just a Theory"?
GRADE-LEVEL USE →	3–12	3–12	3–12	3–12	6–12	9–12	9–12	9–12	6–12	6–12	6–12	9–12
RELATED CONCEPTS ↓												
apparent vs. actual size		X		X								
big bang theory											X	X
galaxies									X	X	X	
objects in the sky	X		X							X		
solar system objects: identity	X	X		X								
speed of light									X		X	
stars: brightness and distance		X	X				X		X			
stars: composition					X		X	X				
stars: evolution						X	X	X				
stars: locations	X	X	X							X		
stars: origin of elements								X				
stars: size		X		X			X					

Teaching and Learning Considerations

This last section opens a window that is hundreds of millions of times larger in scope that the previous section about the solar system. We begin with a series of probes about the "near" universe—that is, within our Milky Way galaxy. The first eight probes concern the nature of stars, how they are distributed in space, the size and composition of stars, how stars change over time, and how they seed the universe with elements that make up our own bodies.

Probe 42, "Seeing Into the Past," plays a dual role. It provides information on your students' understanding of the nature of light and on the use of light as a unit of measure that works on the huge scale of stars and galaxies.

Probe 43, "What Is the Milky Way?" also has a dual role. It elicits students' understanding of the solar system's location within our galaxy, and it also helps you determine if students understand the connection between the Milky Way that is visible in a dark sky and the beautiful telescopic images of spiral galaxies that students can download from the web.

The last two probes in this section concern the largest scale of all—the big bang theory of the origin of the universe. These probes elicit your students' understanding of the most fundamental aspects of this theory, and their understanding of the nature of a theory and its use in science. The last probe, "Is the Big Bang 'Just a Theory'?" could be used to build a bridge from astronomy to all of the other sciences, where students could compare the big bang theory with other major theories such as the kinetic molecular theory in physical science, the theory of biological evolution in life science, and the theory of plate tectonics in Earth science.

Related Curriculum Topic Study Guides*

Origin and Evolution of the Universe
Stars and Galaxies
The Universe

*These guides are found in Keeley, P. 2005. *Science Curriculum Topic Study: Bridging the Gap Between Standards and Practice.* Thousand Oaks, CA: Corwin Press and Arlington, VA: NSTA Press. Each Curriculum Topic Study Guide provides a process to help the reader (1) identify adult content knowledge, (2) consider instructional implications, (3) identify concepts and specific ideas, (4) examine research on learning, (5) examine coherency and articulation, and (6) clarify state standards and district curriculum.

Related NSTA and Other Resources

NSTA Press Books

American Association for the Advancement of Science (AAAS). 2001. *Atlas of science literacy.* Vol. 1. (See "Stars" map, pp. 46–47 and "Galaxies and the Universe" pp. 48–49.) Washington, DC: AAAS.

Holt, G., and N. West. 2011. *Project Earth science: Astronomy.* 2nd ed. Arlington, VA: NSTA Press.

Keeley, P., F. Eberle, and C. Dorsey. 2008. *Uncovering student ideas in science, vol. 3: Another 25 formative assessment probes.* (See "Is It a Theory," pp. 83–91, and "Where Do Stars Go?" pp. 191–196.) Arlington, VA: NSTA Press.

Keeley, P., F. Eberle, and J. Tugel. 2007. *Uncovering student ideas in science, vol. 2: 25 more formative assessment probes.* (See "Emmy's Moon and Stars," pp. 177–183, and "Objects in the Sky," pp. 185–190.) Arlington, VA: NSTA Press.

Keeley, P. and R. Harrington. 2010. *Uncovering student ideas in physical science, vol. 1: 45 new formative assessment probes.* Arlington, VA: NSTA Press.

NSTA Journal Articles

Brunsell, E., and J. Marcks. 2007. Teaching for conceptual change in space science. *Science Scope* 30 (9): 20–23.

Chaisson, E. 2005. Trekking across the science boundaries. *The Science Teacher* 72 (2): 26–28.

Keeley, P. 2011. Formative assessment probes: Where are the stars? *Science and Children* 49 (1): 32–34.

Larson, B. 2009. Science shorts: Astronomies of scale. *Science and Children* 47 (2): 54–56.

Murphy, E., and R. Bell. 2005. How far are the stars? *The Science Teacher* 72 (2): 38–43.

Riddle, B. 2003. Scope on the skies: Scintillating stars. *Science Scope* 26 (5): 56–57.

Riddle, B. 2008. Scope on the skies: Deep sky objects. *Science Scope* 31 (6): 60–62.

Riddle, B. 2009. Scope on the skies: The new Milky Way galaxy. *Science Scope* 32 (6): 72–79.

Robertson, W. 2008. Science 101: How do we know the universe is expanding, and what exactly does that mean? *Science and Children* 46 (1): 62–65.

Scheider, W. 2005. Idea bank: A big bang lab. *The Science Teacher* 72 (7): 74–75.

Young, D. 2005. Our cosmic connection. *The Science Teacher* 72 (2): 29–31.

NSTA Learning Center Resources

NSTA SciGuides

http://learningcenter.nsta.org/products/sciguides.aspx

Gravity and Orbits

NSTA SciPacks

http://learningcenter.nsta.org/products/scipacks.aspx

The Universe

NSTA Science Objects

http://learningcenter.nsta.org/products/science_objects.aspx

Universe: How Do We Know What We Know?

Universe: Sun as a Star

Universe: The Universe Beyond Our Solar System

Universe: Birth, Life, and Death of Stars

Universe: The Origin and Evolution of the Universe

Other Resources

Fraknoi, A., ed. 2011. *Universe at your fingertips 2.0.* San Francisco: Astronomical Society of the Pacific. Available at *www.astrosociety.org/uayfl/index.html*

GEMS Space Science Sequence. Available from Carolina Biological Supply Company at *www.carolinacurriculum.com/GEMS*

Shooting Star

Three girls are having a sleepover. It's a clear summer evening and the stars are out. All of a sudden one of them points to a place in the sky and says, "Look! It's a shooting star!" The girls saw a bright streak in the sky, and then it was gone. They each wondered what a shooting star actually is. This is what they said:

Kaitlin: "I think it is a comet with a bright, glowing tail."

Maria: "I think it is a star falling out of the sky."

Tiara: "I think it is a rock from space burning up."

Whom do you agree with the most? ____________ Explain why you agree with one friend and not the others.

____

____

____

____

____

Shooting Star

Teacher Notes

Purpose

The purpose of this assessment probe is to elicit students' ideas about objects in the night sky. The probe is designed to uncover students' understanding of the nature of meteors, comets, and stars.

Related Concepts

Objects in the sky
Solar system objects: identity
Stars: locations

Explanation

Tiara has the best answer: "I think it is a rock from space burning up." The girls have just witnessed a meteor—a large rock burning up from friction within the atmosphere as it streaks to Earth at a very high speed. Comets are much larger bodies, usually consisting mostly of ice that turns to vapor as they near the Sun. A comet leaves a long tail that streams away from the head of the comet in the opposite direction from the Sun, so we see the comet while it is still out in space. Comets appear to move very slowly and do not streak across the sky like meteors. Stars are thousands of times more massive than Earth, very hot, and very, very far away. If a star were to come close to Earth, we would be vaporized long before it touched our atmosphere.

Administering the Probe

This probe is best used with students in upper elementary or middle school grades to launch a unit about objects in space. It can be used to check for misconceptions at the high school level when students are learning about stars and objects outside of our solar system. Begin

the probe by asking students if anyone has ever seen a shooting star.

Related Ideas in *Benchmarks for Science Literacy* (AAAS 2009)

6–8 The Universe

- ★ Many chunks of rock orbit the Sun. Those that meet the Earth glow and disintegrate from friction as they plunge through the atmosphere—and sometimes impact the ground.

Related Ideas in *National Science Education Standards* (NRC 1996)

5–8 Earth in the Solar System

- The Earth is the third planet from the Sun in a system that includes the Moon, the Sun, eight other planets and their moons, and smaller objects such as asteroids and comets. *[Note: This standard was written before Pluto was reclassified.]*

Related Research

- Neil Comins, in his book *Heavenly Errors: Misconceptions About the Real Nature of the Universe* (2003) and companion website (*www.physics.umaine.edu/ncomins*) listed common misconceptions from his college astronomy students. One of these misconceptions is that shooting stars or falling stars are actual stars whizzing across the universe or falling through the sky.
- The term *shooting star* or *falling star* applied to a meteor may well be a common source of the misconception that stars are small objects within the solar system that occasionally fall from the sky. Agan (2004) interviewed high school and college students concerning their ideas about stars. In response to the open-ended questions "Do stars change over time?" and "Is there anything left after a star stops shining?" Those interviewed included eight high school freshmen (ages 14–15) who had received minimal astronomy instruction in an Earth science class; four high school juniors and seniors (ages 16–18) who were completing a semester-long astronomy course; and five college students (ages 18–19) who had not received any astronomy instruction in high school or college. 12% of the high school students taking an Earth science class replied that stars turn into shooting stars. None of the high school students taking astronomy or the college students expressed that misconception.

Suggestions for Instruction and Assessment

- To counter misconceptions that may result from common references to meteors as shooting stars or falling stars, it may be useful to point out to students after they discuss their responses to the probe that the way we use words in our everyday language is not always consistent with their scientific use.
- After the students write their own answers to the probe, list their different ideas on chart paper or on the board, plus any other ideas that students suggest. Next to the list create a "Pros" column and a "Cons" column to list reasons for and against that idea. As students learn more about celestial objects and about gravity, they can update the list and develop a scientific understanding of meteors.
- Students can research the history of scientific understanding of meteors at reputable websites such as this one created by the American Museum of Natural History: *www.amnh.org/exhibitions/permanent/*

★ Indicates a strong match between the ideas elicited by the probe and a national standard's learning goal.

meteorites/what/history.php; here they will find that there was a debate among scientists about the origin of the odd rocks known as meteorites. The debate was resolved in the early 1800s when people documented their observations that these rocks actually fell from space. Eventually scientists developed the following definitions to clarify their growing understanding of the phenomenon:

- *Meteoroid:* a small rocky or metallic object in space (large objects are called asteroids)
- *Meteor:* a meteoroid that is falling through Earth's atmosphere, leaving a bright trail as it is heated from friction with the atmosphere
- *Meteorite:* the remnant of a meteor after it has fallen to Earth

However, teachers are cautioned to develop a conceptual understanding of these objects before introducing the vocabulary; otherwise students may memorize definitions with little conceptual understanding.

- Look for an opportunity to observe a *meteor shower* with your students. On a clear evening away from city lights when there is no Moon in the sky, students are likely to see a bright meteor every few minutes during a meteor shower. These events, which last a day or two, occur when Earth passes through a region where there are many meteoroids. These accumulations of debris are closely associated with comets, which gradually shed material as they enter the inner solar system (Jenniskens 2006). Meteor showers can be predicted and are listed on the International Meteor Organization website: *www.imo.net*
- The origin of meteors should be revisited at the high school level, and this probe can be used to diagnose possible misconceptions at the start of a unit. High school students can engage much more deeply into the meteor phenomenon, studying such topics as the use of meteorites to determine the age of the solar system. Students may find particularly interesting the strange case of the meteorite named Allan Hills 84001, which some scientists think may bear evidence of bacterial life on Mars.

References

Agan, L. 2004. Stellar ideas: Exploring students' understanding of stars. *Astronomy Education Review* 3 (1): 77–97. *http://aer.aas.org/resource/1/aerscz/v3/i1/p77_s1*

American Association for the Advancement of Science (AAAS). 2009. Benchmarks for science literacy online. *www.project2061.org/publications/bsl/online*

Comins, N. F. 2003. *Heavenly errors: Misconceptions about the real nature of the universe.* New York: Columbia University Press.

Jenniskens P. 2006. *Meteor showers and their parent comets.* Cambridge, U.K.: Cambridge University Press.

National Research Council (NRC). 1996. *National science education standards.* Washington, DC: National Academies Press.

Is the Sun a Star?

Six students were talking about the Sun. They had several different ideas. Here is what they said:

Sam: "Our Sun is just an average star made up of hot glowing gas."

Juanita: "Our Sun is a huge ball of hot, glowing gas, but it is too close to us to be a star."

Tillie: "Our Sun is one of the many stars found inside our solar system."

Skeeter: "Our Sun was once a planet but is now a large burning ball."

Maggie: "Our Sun is a major star. The other stars in the sky are sparks broken off from our Sun."

Diamond: "Our Sun is a sun. It's not a star. Suns are different from stars."

Which student do you agree with the most? ______________ Explain why you agree.

____

____

____

Is the Sun a Star?

Teacher Notes

Purpose

The purpose of this assessment probe is to elicit students' ideas about the Sun and stars. The probe is designed to uncover students' ideas about the nature of stars and their recognition that the Sun is an average star, but much closer to us than other stars.

Related Concepts

Apparent vs. actual size
Solar system objects: identity
Stars: brightness and distance, location, size

Explanation

Sam has the best idea: "Our Sun is just an average star made up of hot glowing gas." The Sun is an average star. Some stars are much bigger and brighter than the Sun, and some are smaller and dimmer. The Sun looks much bigger and brighter because it is thousands of times closer to us than other stars.

Administering the Probe

The probe is primarily designed for students in upper elementary grades or middle school, although it can be used with students in high school as well, both to learn about students' current thinking and to spark conversation as an introduction to a unit on stars. If students aren't sure what is meant by an average star, explain that *average* means average size and brightness.

Related Ideas in *Benchmarks for Science Literacy* (AAAS 2009)

3–5 The Universe

★ Stars are like the Sun, some being smaller and some larger, but so far away that they look like points of light.

6–8 The Universe

★ The Sun is a medium-sized star located near the edge of a disc-shaped galaxy of stars, part of which can be seen as a glowing band of light that spans the sky on a very clear night.

★ The universe contains many billions of galaxies, and each galaxy contains many billions of stars. To the naked eye, even the closest of these galaxies is no more than a dim, fuzzy spot.

- The Sun is many thousands of times closer to the Earth than any other star. Light from the Sun takes a few minutes to reach the Earth, but light from the next nearest star takes a few years to arrive. The trip to that star would take the fastest rocket thousands of years.
- Some distant galaxies are so far away that their light takes several billion years to reach the Earth. People on Earth, therefore, see them as they were that long ago in the past.

Related Ideas in *National Science Education Standards* (NRC 1996)

K–4 Objects in the Sky

- The Sun, Moon, stars, clouds, birds, and airplanes all have properties, locations, and movements that can be observed and described.

5–8 Earth in the Solar System

★ The Sun, an average star, is the central and largest body in the solar system.

Related Research

- Colombo, Aroca, and Silva (2010) administered a questionnaire and interviewed 137 students ages 10–11 during a visit to a college observatory in Brazil. Although the students learned some information about the solar system during the visit, they were not able to develop a comprehensive view of the solar system. For example, 87% of the students thought that the Sun is a star, but only 18% thought the Sun was the only star in the solar system.
- Agan (2004) interviewed high school and college students concerning their ideas about stars. Those interviewed included eight high school freshmen (ages 14–15) who had received minimal astronomy instruction in an Earth science class; four high school juniors and seniors (ages 16–18) who were completing a semester-long astronomy course; and five college students (ages 18–19) who had not received any astronomy instruction in high school or college. In response to the question, "What's the closest star to Earth?" none of the Earth science students identified the Sun as the closest star to Earth; three of the five college students identified the Sun as closest to Earth, and all but one of the high school astronomy students responded that the Sun is the closest star to Earth.
- Lightman, Miller, and Leadbeater (1987) conducted a telephone survey of adults (age 18 and over) and found that 55% of adults correctly identified the Sun as a star and 25% said the Sun is a planet. The remaining adults had a different explanation or no explanation.

★ Indicates a strong match between the ideas elicited by the probe and a national standard's learning goal.

Suggestions for Instruction and Assessment

- At the upper elementary level the primary focus should be on the Earth-Sun-Moon system and how their relative movements result in our experience of everyday phenomena such as the apparent motion of the Sun's daily path in the sky, the day-night cycle, and phases of the Moon. This is also time to introduce the idea that the Sun is a star and all of the stars are suns. However, do not expect that all students will accept the idea, since the Sun appears so much bigger and brighter than the distant stars.
- The research studies referenced earlier in this probe suggest that students can learn the fact that "the Sun is a star" but may not realize the full implications of that statement, because many still think that there are many stars in the solar system. Helping students develop a realistic mental model of the solar system that contains just one star should be a key goal for middle school science.
- One approach, appropriate for middle school students, is to present new information about planetary systems that have been discovered around other stars. Students could begin by researching extrasolar planets at a website such as *http://planetquest.jpl.nasa.gov/*. Having students draw what a distant planetary system might look like based on actual scientists' reports may help students recognize that each planetary system has a central star—just as our own solar system has one star, sometimes called by its Roman name, *Sol*. In some systems there are two (or even more stars) at the center.
- Another approach, appropriate for middle and high school levels, is to have students research the history of the idea that the Sun is a star and that all of the stars in the sky are distant suns. Although many early astronomers thought the stars were probably distant suns, two important lines of evidence were needed to convince everyone that it was true. One of these pieces of evidence was the first actual measurement of distance to a star showing that the stars are thousands of times farther away than the Sun. This discovery is discussed at the following website: *http://cosmology.carnegiescience.edu/timeline/1838*. The other line of evidence involved breaking starlight up into its various colors with a spectroscope, showing that the stars are made of the same elements (mostly hydrogen) as the Sun. That discovery is discussed at the following website: *http://cosmology.carnegiescience.edu/timeline/1861*.

References

Agan, L. 2004. Stellar ideas: Exploring students' understanding of stars. *Astronomy Education Review* 3 (1): 77–97. *http://aer.aas.org/resource/1/aerscz/v3/i1/p77_s1*.

American Association for the Advancement of Science (AAAS). 2009. Benchmarks for science literacy online. *www.project2061.org/publications/bsl/online*

Colombo, P., Jr., S. Aroca, and C. Silva. 2010. Daytime school guided visits to an astronomical observatory in Brazil. *Astronomy Education Review* 9 (1). *http://aer.aas.org/resource/1/aerscz/v9/i1/p010113_s1*

Lightman, A. P., J. D. Miller, and B. J. Leadbeater. 1987. Contemporary cosmological beliefs. In *Second International Seminar on Misconceptions and Educational Strategies in Science and Mathematics*, ed. J. D. Novak, 309–321. Ithaca, NY: Cornell University Press.

National Research Council (NRC). 1996. *National science education standards.* Washington, DC: National Academies Press.

Where Are the Stars in Orion?

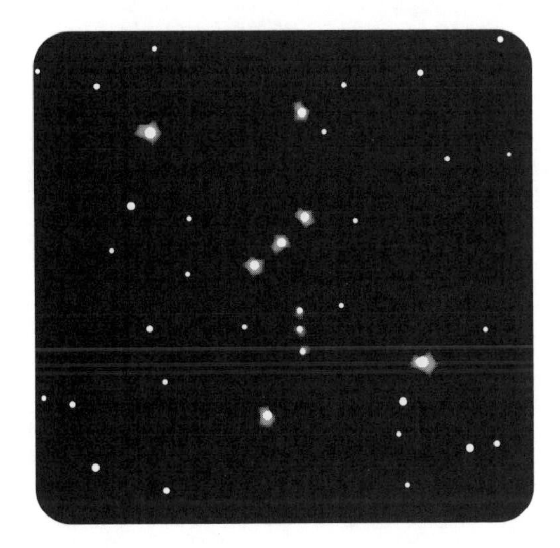

Orion is a constellation—a group of stars in the sky. Circle the statement you think best describes the stars in Orion.

A The stars of Orion are close together in space.

B The stars in Orion orbit the Sun, just like the planets.

C The brightest stars in Orion are the ones that are closest to us.

D You can't tell if the brightest stars in Orion are really brighter than the others, or if they are just closer to us.

E The stars in Orion are all the same distance from us. That is why we see the shape of the constellation.

F The stars are arranged in the constellation Orion for only part of a year. The rest of the year they change their pattern and form different constellations.

Explain your thinking. Describe what you know about stars that form constellations

____

____

Where Are the Stars in Orion?

Teacher Notes

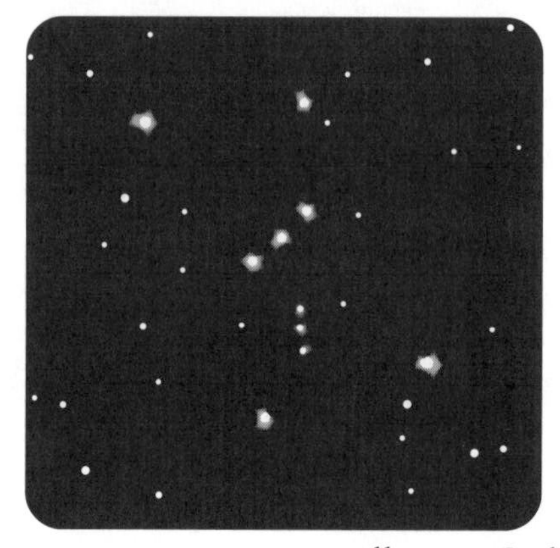

Purpose

The purpose of this assessment probe is to elicit students' understanding of stars in constellations. The probe is designed to find out whether students recognize that stars are distributed in space, that some stars are brighter than others, and that although two stars may appear next to each other with the same brightness, one might be much farther away than the other.

Related Concepts

Objects in the sky
Stars: locations, brightness and distance

Explanation

The best response is D: "You can't tell if the brightest stars in Orion are really brighter than the others, or if they are just closer to us." To an observer, constellations look "flat." It appears as if the stars are all at the same distance from us, with some appearing to be brighter than others. People tend to refer to bright stars as big stars and faint ones as small stars. Or bright stars are considered to be closer and dim stars farther away. But in reality, the opposite could also be true. The bright stars could be farther away than the dimmer ones. And the ones that appear dim could have the same or even more brightness as ones that appear bright, it's just that they may be farther away. You can't tell if one star is actually brighter than the other by observing them with the naked eye. Orion has some of the brightest stars we can see from Earth when we look into the night sky. The stars we see in the Orion constellation range in distances from 243 to 1,360 light-years. Consider just two of the brightest:

- **Rigel,** which marks Orion's left foot, is the brightest in the constellation with a magnitude of 0.1. (*Magnitude* describes brightness. The lower the number, the brighter the star.) Rigel is 777 light-years away and about 55,000 times brighter than the Sun.
- **Bellatrix,** which marks Orion's left shoulder, is closer at 243 light-years but fainter at

magnitude 1.6, and 6,000 times as bright as our Sun.

So you really can't tell by looking if a star in a constellation is more or less luminous (bright) than another star, since you can't tell if it is closer or farther away.

Administering the Probe

The probe can be used with students from elementary school through high school. You might begin the probe by asking students if anyone has ever seen the constellation Orion.

Related Ideas in *Benchmarks for Science Literacy* (AAAS 2009)

3–5 The Universe

- The patterns of stars in the sky stay the same, although they appear to move across the sky nightly, and different stars can be seen in different seasons.
- Stars are like the Sun, some being smaller and some larger, but so far away that they look like points of light.

6–8 The Universe

- The Sun is many thousands of times closer to the Earth than any other star. Light from the Sun takes a few minutes to reach the Earth, but light from the next nearest star takes a few years to arrive. The trip to that star would take the fastest rocket thousands of years.

Related Ideas in *National Science Education Standards* (NRC 1996)

K–4 Objects in the Sky

★ The Sun, Moon, stars, clouds, birds, and airplanes all have properties, locations, and movements that can be observed and described.

Related Research

- Sadler (1998) developed a written test to measure high school students' understanding of astronomy concepts and gave it to 1,250 students in grades 8–12 who were taking an Earth science or astronomy course. One of the questions was as follows:

  The Big Dipper would have a noticeably different shape to the unaided eye:
  A. If viewed from another star.
  B. If viewed from Pluto.
  C. If you looked at it a year from now.
  D. If you viewed it from China.
  E. Never, it would always look the same.
  Only 28% of the students correctly answered A, that the Big Dipper would appear to have a different shape if viewed from another star. To answer the question correctly, students would need to understand that the stars in the Big Dipper are very far away (not in the solar system).

- Dussault (1999) reported the results of a survey of 257 visitors to the National Air and Space Museum in Washington, D.C., who were asked to name things found in the solar system. As expected, 82% named planets, but, surprisingly, 41% named stars as a component of the solar system.
- Colombo, Aroca, and Silva (2010) administered a questionnaire and interviewed 137 students, ages 10–11, during a visit to a college observatory in Brazil. Although the students learned some information about the solar system during the visit, they were not able to develop a comprehensive view of the solar system. For example, 87% of the students responded that the Sun is a star, but only 18% thought the Sun was the only star in the solar system. Consider

★ Indicates a strong match between the ideas elicited by the probe and a national standard's learning goal.

the following interview with a 10-year-old student:

Interviewer: Is the Sun a star?
Student: Yes.
Interviewer: Is it the only star in the solar system?
Student: No.
Interviewer: So how many stars are there in the solar system?

Suggestions for Instruction and Assessment

- The idea that stars that appear together in the sky are not necessarily near each other is not easy to accept. An excellent elementary school activity is to build a physical 3-D model of a constellation showing that when viewed from one position, the stars have the familiar constellation shape, but when viewed from a different angle (as though the observer were traveling to a different star) they take on a different shape. Activity directions based on actual star distances is included in *Universe at Your Fingertips 2.0*, a packet of activities from the Astronomical Society of the Pacific (Fraknoi 2011; see Activity F7. 3-D Constellations). Astronomy simulations at the University of Nebraska–Lincoln can also be used with the 3-D model: *http://astro.unl.edu/classaction/animations/coordsmotion/bigdipper.html*
- The activity described above is also appropriate for middle school students. However, you may provide additional learning opportunities related to how astronomers determine the distances to stars using the parallax method, which means measuring the distance to an object by observing it from two different positions. To demonstrate the essential concept, have each student hold up a thumb about a foot from his or her face and blink with one eye and then the other. The students should notice how the thumb seems to "jump" back and forth as viewed against objects in the distance. Next have each student hold a thumb up at arm's length and do the same blinking action. They will see that the thumb still seems to "jump" but not as far. Ask the students if they can explain why that happens. (Blinking changes the angle from which you see your thumb.) Astronomers use the same method to measure the distance to a star by measuring how far a nearby star appears to "jump" back and forth against background stars when viewed from one side of the Sun and, six months later, from the other side of the Sun.
- Students can learn about the first use of the parallax method to measure the distance to a star at *http://cosmology.carnegiescience.edu/timeline/1861.*
- If the probe shows that high school students are not familiar with the idea that stars are widely distributed in space and have different luminosities, students can view one of the online 3-D maps of Orion or other constellations. For example, they can view an excellent simulated "fly by" of the stars in Orion on YouTube at *www.youtube.com/watch?v=ik013dMQ5ck.*

References

American Association for the Advancement of Science (AAAS). 2009. Benchmarks for science literacy online. *www.project2061.org/publications/bsl/online*

Colombo, P., Jr., S. Aroca, and C. Silva. 2010. Daytime school guided visits to an astronomical observatory in Brazil. *Astronomy Education Review* 9 (1). *http://aer.aas.org/resource/1/aerscz/v9/i1/p010113_s1*

Dussault, M. 1999. How do visitors understand the universe? Studies yield information on planning exhibitions and programs. *ASTC Newsletter* (May/June): 9–11.

Fraknoi, A., ed. 2011. *Universe at your fingertips 2.0.* San Francisco: Astronomical Society of the Pacific. Available at *www.astrosociety.org/uayf/index.html*

National Research Council (NRC). 1996. *National science education standards.* Washington, DC: National Academies Press.

Sadler, P. M. 1998. Psychometric models of student conceptions in science: Reconciling qualitative studies and distracter-driven assessment instruments. *Journal of Research in Science Teaching* 35 (3): 265–296.

Which Is Bigger?

Ms. Moody gave her students a piece of paper with the names of five objects. She then asked the students to write numbers under the names to show their sizes. The directions were: Write a 1 under the name of the smallest object and a 5 under the name of the biggest object.

Moon	Saturn	Earth	Star	Sun
_____	_____	_____	_____	_____

One of the students said, "This is not fair!" It's not possible to tell which is the biggest object!"

Do you agree, partially agree, or disagree with the student? ____________________

Explain how you would rank the size of these objects and why.

____

____

____

Which Is Bigger?

Teacher Notes

Purpose

The purpose of this assessment probe is to elicit students' ideas about the size of objects in the sky. The probe is designed to see if students recognize that the Sun and other stars are much bigger than the Earth, Moon, or any of the planets but that stars vary in size compared with our Sun.

Related Concepts

Apparent vs. actual size
Solar system objects: identity
Stars: size

Explanation

The best response is to partially agree with the student. Since the Sun is a medium-size star, it could be bigger or smaller than a more distant star. So the best way to answer the question is to say that the Moon is the smallest (1); Earth is larger (2); and Saturn is larger than Earth (3). The Sun and stars are both bigger than the other three objects, but a star could be smaller or larger than our Sun. It would be okay to write the number (4) or (5) for both the Sun and the star, or to leave those spaces blank and explain why.

Administering the Probe

The probe is primarily designed for students in upper elementary grades or middle school, but it could be useful for high school students as well because research studies show that many high school students and adults are confused about the relative sizes and distance of celestial objects. Make sure students know they have three choices: they can agree with the student who says the question is not fair; they can partially agree since some of the objects can be ranked, but not all of them; or they can disagree, and rank all of the objects according to size. If students know that stars are suns, explain that the sun in this probe is our Sun and the star is a star other than our Sun.

Related Ideas in *Benchmarks for Science Literacy* (AAAS 2009)

3–5 The Universe

- Stars are like the Sun, some being smaller and some larger, but so far away that they look like points of light.

6–8 The Universe

- Nine planets of very different size, composition, and surface features move around the Sun in nearly circular orbits. *[Note: This benchmark was written before Pluto was reclassified.]*
- The Sun is a medium-sized star located near the edge of a disc-shaped galaxy of stars, part of which can be seen as a glowing band of light that spans the sky on a very clear night.

Related Ideas in *National Science Education Standards* (NRC 1996)

K–4 Objects in the Sky

- The Sun, Moon, stars, clouds, birds, and airplanes all have properties, locations, and movements that can be observed and described.

5–8 Earth in the Solar System

- The Sun, an average star, is the central and largest body in the solar system.

Related Research

- Sharp (1996) reported on interviews with 42 children, ages 10–11, who had learned about the solar system through England's national curriculum. When asked about the relative sizes of the Earth, the Moon, and the Sun, most of the students (62%) were aware that the Sun is bigger than the Earth, which is bigger than the Moon. However, fewer than half of the students gave one of two correct answers that stars are bigger than the Earth, the Moon, and the Sun (36%) or that stars are various sizes (10%).
- Dussault (1999) reported on an activity used during interviews with visitors to a science center in preparation for a new museum exhibit. Visitors were given several cards on which were placed images of several celestial objects: the Sun, the Moon, the Hubble Space Telescope, the Pleiades star cluster, Saturn, a spiral galaxy, and the Hubble Deep Field galaxies. Visitors were asked to place the images in order of size and then distance. It was found that "stars and galaxies are often placed in the solar system, or their location is guessed at" (p. 11). These results showed that few adults are aware of the vast differences in size among celestial objects.

Suggestions for Instruction and Assessment

- It's not surprising that many students are confused about the relative sizes of objects in the solar system, given the way that these objects are depicted in books. For example, any illustration involving planetary orbits around the Sun must show the Sun at least 100 times smaller than it would be if using the same scale for the planets, or a diagram would be impossible to draw.
- Upper elementary grades or middle school is a good time for students to create a model of the solar system on school grounds using the same scale for both size and distance. Several such models are available on the internet, or as part of a rich packet of astronomy activities from the Astronomical Society of the Pacific (Fraknoi 2011; see "Activity D4. The Earth as a Peppercorn";

this activity can also be found at *www.noao.edu/education/peppercorn/pcmain.html*).

- If you use the "Peppercorn" activity, follow up the activity with this probe. The probe can help you see what students learned from the activity. Allow discussion if there are still differences of opinion, and suggest that students conduct research on the web if they still disagree.
- Research indicates that many students at the high school and even college level are confused about relative sizes of celestial objects. An advanced version of the "Peppercorn" activity is to assign the students to create their own scale model, using the same scale for size and distance. Recommend that they start with a familiar object to represent the Sun and then work out the sizes of the other elements of the model, deciding what objects would be the right size to represent the other components of the solar system and how far they should be placed from the Sun. Then challenge them to add to their model Alpha Centauri, the closest visible star to the Sun.

References

American Association for the Advancement of Science (AAAS). 2009. Benchmarks for science literacy online. *www.project2061.org/publications/bsl/online*

Dussault, M. 1999. How do visitors understand the universe? Studies yield information on planning exhibitions and programs. *ASTC Newsletter* (May/June): 9–11.

Fraknoi, A., ed. 2011. *Universe at your fingertips 2.0.* San Francisco: Astronomical Society of the Pacific. Available at *www.astrosociety.org/uayf/index.html*

National Research Council (NRC). 1996. *National science education standards.* Washington, DC: National Academies Press.

Sadler, P. M. 1992. The initial knowledge state of high school astronomy students. Doctoral diss., Graduate School of Education, Harvard University.

Sharp, J. 1996. Children's astronomical beliefs: A preliminary study of year 6 children in Southwest England. *International Journal of Science Education* 18 (6): 685–712.

What Are Stars Made Of?

Mr. Willard asked his students how astronomers figure out what stars are made of. This is what his students said:

Igor: "I think it's impossible to know what stars are made of."

Bonnie: "I think astronauts are able to take samples that astronomers can look at to see what they are made of."

Olympia: "Astronomers can tell what our Sun is made from, and that tells them what stars are made of."

Sophia: "I think astronomers examine the light given off by a star to see what it's made of."

Aida: "When shooting stars fall to Earth, some pieces are left. Astronomers take them to the lab to see what they're made of."

Which student do you think has the best answer to Mr. Willard's question?

____

Explain why you think that is the best answer.

____

What Are Stars Made Of?

Teacher Notes

Purpose

The purpose of this assessment probe is to elicit students' ideas about how astronomers investigate the composition of stars. The probe is designed to see if students recognize that the colors from the light given off by a star (its spectrum) reveal information about the composition of a star.

Related Concepts

Stars: composition

Explanation

Sophia has the best answer: "I think astronomers examine the light given off by a star to see what it's made of." Light radiates from stars. Each element can be identified by its pattern of colors and absorption lines (called a spectrum) that appear when light from the star passes through an instrument called a spectroscope. Astronomers compare the spectrum of stars with spectra that they produce in the laboratory to determine what the stars are made of.

Administering the Probe

This probe is designed for students in middle school or high school, after they learn about how to divide light into its distinctive colors or "spectra."

Related Ideas in *Benchmarks for Science Literacy* (AAAS 2009)

6–8 Motion

- Light from the Sun is made up of a mixture of many different colors of light, even though to the eye the light looks almost white. Other things that give off or reflect light have a different mix of colors.

9–12 The Universe

- The stars differ from each other in size, temperature, and age, but they appear to be made up of the same elements found on

Earth and behave according to the same physical principles.

★ Increasingly sophisticated technology is used to learn about the universe. Visual, radio, and x-ray telescopes collect information from across the entire spectrum of electromagnetic waves; computers handle data and complicated computations to interpret them; space probes send back data and materials from remote parts of the solar system; and accelerators give subatomic particles energies that simulate conditions in the stars and in the early history of the universe before stars formed.

Related Ideas in *National Science Education Standards* (NRC 1996)

9–12 Interactions of Energy and Matter

★ Each kind of atom or molecule can gain or lose energy in particular discrete amounts and thus can absorb and emit light only at wavelengths corresponding to these amounts. These wavelengths can be used to identify the substance.

Related Research

- Sharp (1996) reported on interviews with 42 children, ages 10–11, who had learned about the solar system through England's national curriculum. When the children were asked what stars are, some said they are "like the Sun," or "a ball of gases like the Sun." Other responses included that stars are "small planets, couldn't stand on them, though, they're too far away"; that "stars are little things, but far away, round and with glowing things on them"; that stars are "rocks with gases inside that shine"; and that stars "formed after the Big Bang, they've got poisonous gases inside."
- Agan (2004) interviewed high school and college students concerning their ideas about the stars. Those interviewed included eight high school freshmen, ages 14–15, who had received minimal instruction in astronomy; four high school juniors and seniors, ages 16–18, who were completing a semester-long astronomy course; and five first-year college students, ages 18–19, who had not received any astronomy instruction in high school or college. In response to the question "What is a star?" most of the high school freshmen and all of the college students described the Sun as a "ball of gas," or a "ball of fire." However, the high school students who had just completed a course in astronomy recognized the production of energy as a defining feature of stars.
- Sadler et al. (2010) developed and validated a multiple-choice assessment item test bank for astronomy and space science in which the alternative answers for each question were based on research findings about common misconceptions. Results were reported for a sample of 7,599 students and their 88 teachers on the MOSART (Misconceptions-Oriented Standards-Based Assessment Resources for Teachers) website: *www.cfa.harvard.edu/smgphp/mosart/aboutmosart_2.html*. Items are listed for grades K–4, 5–8, and 9–12. Since each question has five answer choices, the odds of selecting the correct answer by chance are 20%. Two relevant items were as follows:

  Stars begin their lives composed primarily of:
  A. propane
  B. hydrogen
  C. neon
  D. molten rock
  E. uranium
  The correct answer, B, was given by 62%. According to the researchers: "The com-

★ Indicates a strong match between the ideas elicited by the probe and a national standard's learning goal.

position of the Sun may be one reason so many students answered correctly. Very few discuss either the gaseous or solid components listed. No other option received more than 15% of the responses (Option E)."

Which of these elements found on Earth is, or are, also present in the Sun?

A. hydrogen
B. helium
C. neon
D. two of the above
E. all of the above

The correct answer, E, was given by 17%. According to the researchers: "This is a difficult item, as a disproportionate number of students (55%) chose D, presumably thinking of hydrogen and helium, the two most common elements that comprise the Sun."

Suggestions for Instruction and Assessment

- At the elementary level it is sufficient for students to think of the Sun and stars as consisting of an extremely hot glowing gas. It is premature to introduce how a star's spectrum can be used to determine its constituent elements.
- Middle school students should have an opportunity to observe the spectrum of light from various sources. This can be done with a piece of film ruled with thousands of tiny lines, called a diffraction grating. Inexpensive diffraction grating films are available from several sources, and surplus CDs can also be used as diffraction gratings. **[Safety note: Students should not look directly through a diffraction grating at the Sun.]** Safe ways of observing the solar spectrum appear on the following websites:
    - *www.cs.cmu.edu/~zhuxj/astro/html/spectrometer.html*
    - *www.stargazing.net/david/spectroscopy/SimpleNeedleSpectroscope.html*
- Students can also make a simple diffraction grating spectroscope to observe light sources around them. When viewed through a diffraction grating, incandescent sources, such as lightbulbs with filaments and candles, display a "continuous" spectrum in which the colors blend together. Fluorescent bulbs and neon lights display a "line" spectrum in which bright lines of specific colors stand out. The continuous spectrum indicates temperature, with yellow and white being hottest and orange and red coolest. Also, each glowing element has its own distinct pattern of lines, making it possible to determine the elements that make up a star. Some high schools have gas discharge tubes that can be used to examine the spectra of different gases that make up stars. You can find instructions for making a spectroscope at *www.youtube.com/watch?v=YStZk2zANvk.* Students can view the spectra from stars at other websites, such as *http://stars.astro.illinois.edu/sow/spectra.html*, to see how lines are used to determine composition.
- A common use for spectra at the high school level is for students to learn how light is "red shifted" when a star or galaxy is moving toward or away from the observer. It is important for the students to learn to use spectra to determine color and composition first in the lab, before advancing to study how spectral lines are shifted toward the blue when an object is approaching, or toward the red when it is moving away from us.

References

Agan, L. 2004. Stellar ideas: Exploring students' understanding of stars. *Astronomy Education Review* 3 (1): 77–97. *http://aer.aas.org/resource/1/aerscz/v3/i1/p77_s1*

American Association for the Advancement of Science (AAAS). 2009. Benchmarks for science literacy online. *www.project2061.org/publications/bsl/online*

National Research Council (NRC). 1996. *National science education standards.* Washington, DC: National Academies Press.

Sadler, P. M., H. Coyle, J. L. Miller, N. Cook-Smith, M. Dussault, and R. R. Gould. 2010. The Astronomy and Space Science Concept Inventory: Development and validation of assessment instruments aligned with the K–12 National Science Standards. *Astronomy Education Review* 8 (1). *http://aer.aas.org/resource/1/aerscz/v8/i1/p010111_s1.* Questions and findings for each question are available at the MOSART website: *www.cfa.harvard.edu/smg-php/mosart*

Sharp, J. 1996. Children's astronomical beliefs: A preliminary study of year 6 children in Southwest England. *International Journal of Science Education* 18 (6): 685–712.

What Happens to Stars When They Die?

Stars are "born" when they begin to shine, but what happens at the end of their "life"? Put an X next to all of the things you think can happen to stars when they "die."

____**A** Stars never die. They keep glowing forever.

____**B** When stars die they fall to Earth. They're called "shooting stars," or meteors.

____**C** Dying stars get dimmer and dimmer and finally stop glowing.

____**D** Dying stars can explode.

____**E** A dying star can become a black hole.

____**F** Stars can expand and get bigger right before they die.

Explain in your own words what happens to a star at the end of its "life."

____

____

____

What Happens to Stars When They Die?

Teacher Notes

Purpose
The purpose of this assessment probe is to elicit students' ideas about stars' life cycle. The probe is designed to reveal what students think occurs at the end of a star's life.

Related Concepts
Stars: evolution

Explanation
The best answers are C, D, E, and F. These are all things that can happen as a star dies. The fate of any particular star—whether it explodes or becomes dimmer and dimmer and finally stops glowing—depends on its initial mass. In brief:

- A low-mass star like the Sun or smaller stars will swell in size, becoming a red giant. (This will happen to the Sun in about 5 billion years.) Eventually this type of star will expel some of its mass into the surrounding space, and what is left will settle down to become a *dwarf star*, about the size of Earth, and will gradually cool off.
- A high-mass star, more than one and a half times as massive as the Sun, will explode as a supernova, blasting a huge amount of matter into space. What is left will settle into a very small dense body called a neutron star, which is about the size of a large city.
- A giant star, from 10 to 20 times as massive as the Sun, will also explode as a supernova, leaving behind an even smaller and denser body called a black hole. Black holes are so dense that their strong gravitational fields will not even allow light to escape.

Administering the Probe
The probe is primarily designed for high school students. It is intended to gather students' preconceptions and general recollection about the alternative fates of stars of different masses (after instruction), not to assess whether or not they recall the details. The probe can be

extended to ask students: "How can astronomers tell what happens to stars when they die?" Similar questions could be asked about different phases of stellar evolution, such as the birth of stars, their lives on the "main sequence" and periods in which stars become unstable. Since the format of this probe is a justified list, make sure students know they can check off multiple answers.

Although astronomers frequently use terms such as *born, grow, life cycle,* and *die* in connection with stars, stars are clearly not *living* in the biological sense of the word. If not careful, use of such analogies can create misconceptions for students about the biological meaning of these terms. You might explain to students that *born* and *die* used in the context of stars do not have the same meaning as these terms in a life science context.

Related Ideas in *Benchmarks for Science Literacy* (AAAS 2009)

9–12 The Universe

- Stars condensed by gravity out of clouds of molecules of the lightest elements until nuclear fusion of the light elements into heavier ones began to occur. Fusion released great amounts of energy over millions of years.
- Eventually, some stars exploded, producing clouds of heavy elements from which other stars and planets could later condense. The process of star formation and destruction continues.
- Increasingly sophisticated technology is used to learn about the universe. Visual, radio, and x-ray telescopes collect information from across the entire spectrum of electromagnetic waves; computers handle data and complicated computations to interpret them; space probes send back data and materials from remote parts of the solar system; and accelerators give subatomic particles energies that simulate conditions in the stars and in the early history of the universe before stars formed.

Related Ideas in *National Science Education Standards* (NRC 1996)

9–12 The Origin and Evolution of the Universe

- Early in the history of the universe, matter, primarily the light atoms hydrogen and helium, clumped together by gravitational attraction to form countless trillions of stars.

Related Research

- Agan (2004) interviewed high school and college students concerning their ideas about stars. Those interviewed included eight high school freshmen (ages 14–15) who had received minimal astronomy instruction in an Earth science class; four high school juniors and seniors (ages 16–18) who were completing a semester-long astronomy course; and five college students (ages 18–19) who had not received any astronomy instruction in high school or college. In response to the questions "Do stars change over time? Is there anything left after a star stops shining?" the high school students who had nearly completed an Earth science class gave various answers, including that stars go through cycles (30%), burn out or fade away (30%), leave material behind (20%), turn into shooting stars (12%), and merge with other stars (10%). Most of the college students responded that stars burn out or fade away (67%), while some stated that they leave material behind (33%). All of the high school students enrolled in the astronomy course gave some account of stellar evolution and said that stars leave some material behind.

- Brickhouse et al. (2002) investigated the responses of 340 undergraduate students enrolled in an introductory astronomy course that included explicit discussion of scientific theories and their evidentiary basis. Data were gathered from the entire class and from in-depth interviews and focus groups with 19 of the students. The researchers found that how students talked and wrote about the nature of science depended on the particular scientific topic under discussion. For example, when discussing the theory of gravitation, many students did not separate theory from evidence, but rather saw it as fact: "Well gravity I guess really isn't a theory. It's a measure more. Gravity in itself is ... they have a definite measurement for it as 9.8 whatever the measurement is, so I'm not sure I would classify that as a theory. It exists." On the other hand, the students tended to be much more skeptical of the theory of stellar evolution, even though the instructor spent a month developing evidence in support of the prevailing theory. For example, one student explained why it is "still a theory" as follows:

  They don't have all the technical knowledge to get the measurements and everything. The distance and everything of not being there to actually see close up. I mean other than the Hubble telescope. But you still can't see all the mechanics that are happening. Some of it is just their conclusions and may not be accurate.

Suggestions for Instruction and Assessment

- This probe can be combined with "Is It a Theory?" in *Uncovering Student Ideas in Science, Vol. 3: Another 25 Formative Assessment Probes* (Keeley, Eberle, and Dorsey 2008) to establish what students think a theory is before they learn about the theory of stellar evolution.
- At the elementary school level it is not necessary or appropriate to teach stellar evolution.
- At the middle school level students will learn about the birth of the solar system and the idea that the Sun begins to glow when it pulls in enough gas and dust to ignite the nuclear furnace. However, details of stellar evolution should wait until high school.
- Today astronomers have high confidence in a theory for how the Sun and other stars are born, live, and eventually die. This widely accepted theory has become a cornerstone of modern science and should be included in every student's high school curriculum, whether as part of an Earth and space science course, a physics course, or an integrated science course. However, it is important that students not learn the theory of stellar evolution as simple fact, but rather how these ideas came into the mainstream.
- A\ major milestone in unraveling the mystery of stellar evolution was the development of the Hertzsprung-Russell (H-R) diagram, in which the color (and therefore temperature) of a star is plotted versus the brightness of the star. The two astronomers who developed this chart independently did not realize that it would provide the key to unlocking the mystery of stellar evolution. However, it eventually became clear that plotting a large number of stars in this way revealed a pattern that led to the discovery of how stars change as they evolve. Students can use the following applet to read about the history of the H-R diagram, create their own H-R diagrams,

or use online simulations: *www.mhhe.com/physsci/astronomy/applets/Hr/frame.html.*

- Complementary to the H-R diagram was the development of high-speed computers that could model what occurs inside a star over time. As these models were improved, they came to predict how stars would be expected to change their position on the H-R diagram, lending further support to developing theories of stellar evolution. A more complete explanation of how the H-R diagram helped astronomers develop theories of stellar evolution is presented at *www.tim-thompson.com/hr.html.*

References

Agan, L. 2004. Stellar ideas: Exploring students' understanding of stars. *Astronomy Education Review* 3 (1): 77–97. *http://aer.aas.org/resource/1/aerscz/v3/i1/p77_s1*

American Association for the Advancement of Science (AAAS). 2009. Benchmarks for science literacy online. *www.project2061.org/publications/bsl/online*

Brickhouse, N. W., Z. R. Dagher, H. L. Shipman, and W. J. Letts. 2002. Evidence and warrants for belief in a college astronomy course. *Science and Education* 11 (6): 573–588.

Keeley, P., F. Eberle, and C. Dorsey. 2008. *Uncovering student ideas in science, vol. 3: Another 25 formative assessment probes.* Arlington, VA: NSTA Press.

National Research Council (NRC). 1996. *National science education standards.* Washington, DC: National Academies Press.

Do Stars Change?

Two friends were looking up at the stars on an especially clear night. They each had different ideas about what they saw. This is what they said:

Lucille: "I think the stars in the sky never change. The stars we are looking at right now were just as bright and beautiful as in the days of the dinosaurs."

Philippe: "I think the stars change over millions of years. Some may get brighter, others may get dimmer. Some stars may even die, while other stars might be born."

Whom do you agree with the most? _______________ Explain why you agree.

____

____

____

____

____

____

Do Stars Change?

Teacher Notes

Purpose

The purpose of this assessment probe is to elicit students' ideas about what happens to stars over time. The probe is designed to reveal whether students recognize that occasionally new stars are "born," change over time, and eventually "die."

Related Concepts

Stars: brightness and distance, composition, evolution, size

Explanation

Philippe has the best idea. Although most stars have extremely long "lives," new stars are always being "born," stars change in brightness and size, and they eventually "die." The features of a star's life cycle depend on the initial mass of the star. (Although stars are not actually alive, astronomers use terms like *born, die,* and *life cycle* as metaphors to describe how stars evolve.) The most massive stars only survive a few million years before they explode, throwing off much of their mass into interstellar space. Smaller stars like the Sun survive for billions of years. During their life cycles stars become brighter or dimmer and expand or shrink in size. Although no one has lived long enough to observe a single star from its birth to its death, astronomers have confirmed their understanding of how stars evolve by observing many stars at different phases of their life cycles. Astronomers have observed areas in space, such as a region in the constellation Orion, where huge clouds of gas and dust are collapsing under the influence of gravity, until they are dense enough to start glowing as new stars. Astronomers have also observed stars explode.

Administering the Probe

The probe is primarily designed for high school students to elicit ideas before learning about the evolution of stars. The probe addresses only what happens to stars, not the

mechanism that causes stars to glow in the first place and to change over time: nuclear fusion. If it is likely that students have encountered these ideas before, you may wish to add a few probing questions to see what your students think about the source of a star's energy and the mechanisms that trigger changes in a star's size, brightness, or stability.

Related Ideas in *Benchmarks for Science Literacy* (AAAS 2009)

9–12 The Universe

- Stars condensed by gravity out of clouds of molecules of the lightest elements until nuclear fusion of the light elements into heavier ones began to occur. Fusion released great amounts of energy over millions of years.
- ★ Eventually, some stars exploded, producing clouds of heavy elements from which other stars and planets could later condense. The process of star formation and destruction continues.
- Increasingly sophisticated technology is used to learn about the universe. Visual, radio, and x-ray telescopes collect information from across the entire spectrum of electromagnetic waves; computers handle data and complicated computations to interpret them; space probes send back data and materials from remote parts of the solar system; and accelerators give subatomic particles energies that simulate conditions in the stars and in the early history of the universe before stars formed.

Related Ideas in *National Science Education Standards* (NRC 1996)

9–12 The Origin and Evolution of the Universe

- Early in the history of the universe, matter, primarily the light atoms hydrogen and helium, clumped together by gravitational attraction to form countless trillions of stars.
- Stars produce energy from nuclear reactions, primarily the fusion of hydrogen to form helium. These and other processes in stars have led to the formation of all the other elements.

Related Research

- Agan (2004) interviewed high school and college students concerning their ideas about stars. Those interviewed included eight high school freshmen (ages 14–15) who had received minimal astronomy instruction in an Earth science class; four high school juniors and seniors (ages 16–18) who were completing a semester-long astronomy course; and five college students (ages 18–19) who had not received any astronomy instruction in high school or college. In response to the questions "Do stars change over time? Is there anything left after a star stops shining?" the high school students enrolled in an Earth science class gave various answers, including that stars go through cycles (30%), burn out or fade away (30%), leave material behind (20%), turn into shooting stars (12%) and merge with other stars (10%). Most of the college students responded that stars burn out or fade away (67%), while some stated that they leave material behind (33%). All of the high school

★ Indicates a strong match between the ideas elicited by the probe and a national standard's learning goal.

students enrolled in the astronomy course gave some account of stellar evolution and said that stars leave some material behind.

Suggestions for Instruction and Assessment

- This probe can be combined with "Is It a Theory?" in *Uncovering Student Ideas in Science, Vol. 3: Another 25 Formative Assessment Probes* (Keeley, Eberle, and Dorsey 2008) to establish what students think a theory is before they learn about the theory of stellar evolution.
- Today astronomers have high confidence in a theory for how the Sun and other stars are born, live, and eventually die. This widely accepted theory has become a cornerstone of modern science and should be included in every student's high school curriculum, whether as part of an Earth and space science course, a physics course, or an integrated science course. While it is not essential that students learn the details of what happens to stars of different masses, it is important that students come to understand how these ideas came into the mainstream.
- Although astronomers frequently use terms such as *born, die,* and *life cycle* in connection with stars, stars are clearly not *living* in the biological sense of the word. If not careful, use of such analogies can create misconceptions for students about the biological meaning of these terms. You might explain to students that *born* and *die* used in the context of stars do not have the same meaning as these terms in a life science context.
- A major milestone in unraveling the mystery of stellar evolution was the development of the Hertzsprung-Russell (H-R) diagram, in which the color (and therefore temperature) of a star is plotted versus the brightness of the star. The two astronomers who developed this chart independently did not realize that it would provide the key to unlocking the mystery of stellar evolution. However, it eventually became clear that plotting a large number of stars in this way revealed a pattern that led to the discovery of how stars change as they evolve. Students can use the following applet to read about the history of the H-R diagram, create their own H-R diagrams, or use online simulations: *www.mhhe.com/physsci/astronomy/applets/Hr/frame.html*
- Complementary to the H-R diagram was the development of high-speed computers that could model what occurs inside a star over time. As these models were improved, they came to predict how stars would be expected to change their position on the H-R diagram, lending further support to developing theories of stellar evolution. A more complete explanation of how the H-R diagram helped astronomers develop theories of stellar evolution is presented at *www.tim-thompson.com/hr.html*
- An excellent source of information on stellar evolution that is accessible to high school students is "The Lives of Stars" by Andrew Fraknoi, available at *www.pbs.org/seeinginthedark/astronomy-topics/lives-of-stars.html.*

References

Agan, L. 2004. Stellar ideas: Exploring students' understanding of stars. *Astronomy Education Review* 3 (1): 77–97. *http://aer.aas.org/resource/1/aerscz/v3/i1/p77_s1*

American Association for the Advancement of Science (AAAS). 2009. Benchmarks for science literacy online. *www.project2061.org/publications/bsl/online*

Keeley, P., F. Eberle, and C. Dorsey. 2008. *Uncovering student ideas in science, vol. 3: Another 25 formative assessment probes.* Arlington, VA: NSTA Press.

National Research Council (NRC). 1996. *National science education standards.* Washington, DC: National Academies Press.

Are We Made of Star Stuff?

A newspaper article reported that astronomers claim people are made of stardust. What does this mean? Circle the answer that best matches what you think "being made of stardust" means.

A "Being made of stardust" is not intended to be taken literally. It means that nobody really knows where atoms come from, any more than we know what "stardust" is.

B The stars produce dust that falls to Earth. Over millions of years this dust becomes incorporated into living cells.

C Most of the other elements that make up our bodies were formed inside stars or during supernova explosions.

D Our Sun is a star and since humans eat plants or eat animals that eat plants, we are made mostly of stuff that originally came from the Sun.

Explain your thinking. What does "being made of stardust" mean to you?

Are We Made of Star Stuff?

Teacher Notes

Purpose

The purpose of this assessment probe is to elicit students' ideas about stars and the origin of the chemical elements. The probe is designed to reveal their thinking about the role of stars in transforming hydrogen and helium into heavier chemical elements.

Related Concepts

Stars: composition, evolution, origin of elements

Explanation

The best answer is C: Most of the other elements that make up our bodies were formed inside stars or during supernova explosions. Except for a few elements formed during the very early universe (primarily hydrogen and helium, and a small amount of lithium), all of the other elements that make up our planet and our bodies were formed inside stars or during supernova explosions. When a star uses up all of its hydrogen fuel, the star contracts and gets even hotter. The increased temperature makes it possible for helium to fuse into carbon and heavier elements up to iron. The heaviest elements are also formed through fusion, but only in very massive stars during a supernova explosion at the end of the star's life.

The fact that Earth has all 92 natural elements is an indication that a supernova explosion occurred in this part of the galaxy before the solar system formed. Some of the elements that make up our bodies were formed during the big bang, some were formed inside stars as they matured, and still others were formed during a supernova explosion.

Administering the Probe

The probe is primarily designed for high school students. It can be used in astronomy, physics, biology, and chemistry classes.

Related Ideas in *Benchmarks for Science Literacy* (AAAS 2009)

9–12 The Universe

- ★ On the basis of scientific evidence, the universe is estimated to be over ten billion years old. The current theory is that its entire contents expanded explosively from a hot, dense, chaotic mass.
- ★ Stars condensed by gravity out of clouds of molecules of the lightest elements until nuclear fusion of the light elements into heavier ones began to occur. Fusion released great amounts of energy over millions of years.
- ★ Eventually, some stars exploded, producing clouds of heavy elements from which other stars and planets could later condense. The process of star formation and destruction continues.
- ★ Our solar system coalesced out of a giant cloud of gas and debris left in the wake of exploding stars about five billion years ago. Everything in and on the Earth, including living organisms, is made of this material.

Related Ideas in *National Science Education Standards* (NRC 1996)

9–12 The Origin and Evolution of the Universe

- ★ Stars produce energy from nuclear reactions, primarily the fusion of hydrogen to form helium. These and other processes in stars have led to the formation of all the other elements.

Related Research

- Sadler et al. (2010) developed and validated a multiple-choice assessment item test bank for astronomy and space science in which the alternative answers for each question were based on research findings about common misconceptions. Results were reported for a sample of 7,599 students and their 88 teachers on the MOSART (Misconceptions-Oriented Standards-Based Assessment Resources for Teachers) website: *www.cfa.harvard.edu/smgphp/mosart/aboutmosart_2.html*. Items are listed for grades K–4, 5–8, and 9–12. Since each question has five answer choices, the odds of selecting the correct answer by chance are 20%. One item was as follows:

  According to scientists, where were oxygen, carbon, and iron atoms created?
  A. in the big bang
  B. in Earth's core
  C. in the interior of stars
  D. in the spaces between the stars
  E. They weren't created, they were always here.

  The correct answer, C, was given by 24%, which is about the percentage expected by chance. The most popular response was A, given by 40%, indicating that many students think all of the elements were formed in the big bang. Almost 20% of students indicated that oxygen, carbon, and iron were created in the Earth's core (B).

Suggestions for Instruction and Assessment

- Middle school is an appropriate time to develop students' understanding that everything is made of particles—molecules, which in turn are made up of atoms that come in 92 different varieties. They may also learn that some of these elements were formed at the beginning of the universe, and others formed inside stars or during huge stellar explosions. However,

★ Indicates a strong match between the ideas elicited by the probe and a national standard's learning goal.

it is not appropriate to go into the detail about the nuclear reactions that formed the elements.

- High school chemistry or physics class is a good venue to introduce the origin of the chemical elements. The idea can be presented as part of the evolution of stars in an astronomy class, a physics class, or a chemistry class.
- Students should have an opportunity to learn about how scientists came to understand the fusion processes that cause stars to shine, and how those same processes create the elements that make up the visible universe.
- You can extend this probe by assessing your students' understanding of nucleosynthesis—the process of creating new atomic nuclei—at the conclusion of a unit on the evolution of stars.

References

American Association for the Advancement of Science (AAAS). 2009. Benchmarks for science literacy online. *www.project2061.org/publications/bsl/online*

National Research Council (NRC). 1996. *National science education standards.* Washington, DC: National Academies Press.

Sadler, P. M., H. Coyle, J. L. Miller, N. Cook-Smith, M. Dussault, and R. R. Gould. 2010. The Astronomy and Space Science Concept Inventory: Development and validation of assessment instruments aligned with the K–12 National Science Standards. *Astronomy Education Review* 8 (1). *http://aer.aas.org/resource/1/aerscz/v8/i1/p010111_s1.* Questions and findings for each question are available at the MOSART website: *www.cfa.harvard.edu/smgphp/mosart*

Seeing Into the Past

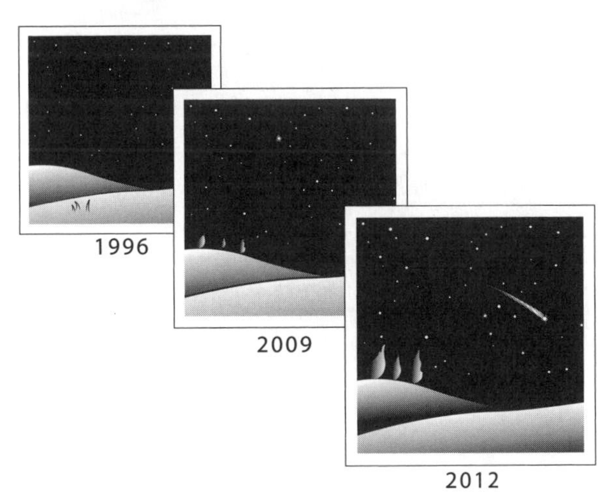

Sara and Ruby are having a lively discussion about "the past." This is what they said:

Sara: "You can never actually see what happened in the past. The best you can do is look at a picture taken by a camera or cell phone or something."

Ruby: "I disagree. You are *always* seeing things in the past because it takes time for light to travel. You are not seeing me as I am *now*. You see me as I *was* a tiny fraction of a second ago since it takes time for the light to travel to your eye."

Sara: "That's silly. Let's suppose it's dark in this room and I don't see you. I turn on the light and there you are! My eyes see you immediately."

Ruby: "The example you gave only seems true because I'm so close to you. It takes light eight minutes to travel from the Sun to Earth. So you don't see the Sun as it is now. You see it as it was eight minutes ago."

Sara: "Huh! If that were true, when I look at the stars I would see them as they were *years* ago. That star could have blown up last year, but I would still be seeing it as it was *before* it blew up! That isn't possible!"

Ruby: "Yes, that's exactly right. We can see into the past."

Circle whom you agree with the most: Ruby Sara

Explain on the back of this page why you agree with one friend and not the other.

Seeing Into the Past

Teacher Notes

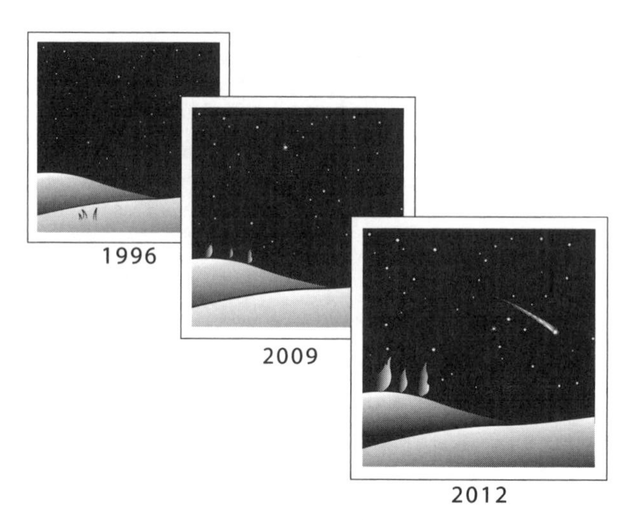

Purpose

The purpose of this assessment probe is to elicit students' ideas about the role of the speed of light in what we see. The probe is designed to see if students recognize that light from far-off objects in space can take many years to reach our eyes, hence we are seeing an object as it looked in the past.

Related Concepts

Galaxies
Speed of light
Stars: brightness and distance

Explanation

Ruby has the best idea. Light travels very quickly—about 300,000 kilometers (186,000 miles) per second in a vacuum—so it appears that it takes "no time at all" for light to travel from a lightbulb to a person's eye. However, it does take a finite amount of time for light to traverse any distance, so we are always seeing into the past. Light slows down a little bit when traveling through transparent media like glass or water, but not enough for us to perceive. However, when it comes to astronomical distances, the travel time of light becomes considerable—about 8 minutes to travel from the Sun to Earth, and 4.3 *years* to travel from the closest visible star (Alpha Centauri) to Earth. The distance that light travels in one year is called a light-year and is used as a unit of measure in astronomy. So the distance between Alpha Centauri and Earth is approximately 4.3 light-years.

Administering the Probe

The probe is designed for middle and high school students. Since this probe is designed in a dialogue format, remind students to follow the conversation all the way through before deciding with whom they most agree. This probe works best in a debate format where students take the side of either Ruby or Sara and engage in similar dialogue and debate.

Related Ideas in *Benchmarks for Science Literacy* (AAAS 2009)

6–8 The Universe

- ★ The Sun is many thousands of times closer to the Earth than any other star. Light from the Sun takes a few minutes to reach the Earth, but light from the next nearest star takes a few years to arrive. The trip to that star would take the fastest rocket thousands of years.
- ★ Some distant galaxies are so far away that their light takes several billion years to reach the Earth. People on Earth, therefore, see them as they were that long ago in the past.

Related Ideas in *National Science Education Standards* (NRC 1996)

5–8 Transfer of Energy

- The Sun loses energy by emitting light. A tiny fraction of that light reaches the Earth, transferring energy from the Sun to the Earth.

Related Research

- Sadler et al. (2010) developed and validated a multiple-choice assessment item test bank for astronomy and space science in which the alternative answers for each question were based on research findings about common misconceptions. Results were reported for a sample of 7,599 students and their 88 teachers on the MOSART (Misconceptions-Oriented Standards-Based Assessment Resources for Teachers) website: *www.cfa.harvard.edu/smgphp/mosart/aboutmosart_2.html.* Items are listed for grades K–4, 5–8, and 9–12. Since each question has five answer choices, the odds of selecting the correct answer by chance are 20%. One item was as follows:

  If the Sun stopped shining right now, the soonest it could be noticed on Earth would be:
  A. a few seconds
  B. a few minutes
  C. a few hours
  D. a few days
  E. a few years
  The correct answer, B, was given by 25%. According to the researchers: "Although the distance between Earth and the Sun is a commonly taught fact, 75% of students do not understand that distance in terms of the time required for light to travel from the Sun to Earth. This lack of understanding may hamper students' understanding of distance and scale within our galaxy and the universe."

Suggestions for Instruction and Assessment

- Middle school is a good time to introduce the idea that it takes time for light to travel. Students have likely noticed that on television news shows it sometimes requires a few seconds for the person being interviewed to respond to a question. That is because it takes time for the signal to go from the TV studio to a satellite, and then back down to the interviewee. The signal generated by the interviewee also takes time to travel back to TV studio.
- Extend the idea of delay in a back-and-forth conversation to consider how long a delay to expect if the person you were talking to were on the Moon (2.4 seconds for a message to go from Earth to the Moon and back again), near the Sun (17 minutes

★ Indicates a strong match between the ideas elicited by the probe and a national standard's learning goal.

for the round-trip), or on a planet near the closest star (8.6 years for the round-trip).

- An excellent project for high school students is to research the history of efforts to measure the speed of light, to explain why early attempts didn't work, and then to develop an explanation—possibly with drawings or a physical model—of the method used by Ole Roemer to finally solve the problem. The following website is useful: *www.speed-light.info/measure/speed_of_light_history.htm.*

References

American Association for the Advancement of Science (AAAS). 2009. Benchmarks for science literacy online. *www.project2061.org/publications/bsl/online*

National Research Council (NRC). 1996. *National science education standards.* Washington, DC: National Academies Press.

Sadler, P. M., H. Coyle, J. L. Miller, N. Cook-Smith, M. Dussault, and R. R. Gould. 2010. The Astronomy and Space Science Concept Inventory: Development and validation of assessment instruments aligned with the K–12 National Science Standards. *Astronomy Education Review* 8 (1). *http://aer.aas.org/resource/1/aerscz/v8/i1/p010111_s1.* Questions and findings for each question are available at the MOSART website: *www.cfa.harvard.edu/smg-php/mosart*

What Is the Milky Way?

The Milky Way is a popular candy bar. The Milky Way is also a part of space. Write YES in front of all of the statements below that describe the Milky Way in space, and NO in front of all the statements that do not describe the Milky Way in space.

_____ **A** a faint band of light across the sky that can be seen on a very dark night

_____ **B** a spiral-shaped group of many stars

_____ **C** a galaxy

_____ **D** our home in the universe

_____ **E** a section of the universe smaller in size than our solar system

_____ **F** a milky-looking area that stretches across the entire universe

_____ **G** a cloud of space dust and debris from planets and stars

_____ **H** a section of the universe that does not include our solar system

_____ **I** billions of stars bound together by gravity

_____ **J** a spiral galaxy in the Virgo cluster of galaxies, 54 million light-years from Earth

In your own words, what is the Milky Way? How does it appear from Earth? How would it appear if we could see it from far out in space?

____

____

What Is the Milky Way?

Teacher Notes

Purpose

The purpose of this assessment probe is to elicit students' ideas about our home galaxy, the Milky Way. The probe is designed to reveal whether students recognize that our solar system is inside the Milky Way, that on clear moonless nights we can see the disc of the Milky Way as a faint band of light across the sky, and that from far out in space our home galaxy would appear as a brightly lit spiral.

Related Concepts

Galaxies
Objects in the sky
Stars: locations

Explanation

The best answers are A, B, C, D, and I. Following is a comment on each of the possible responses.

A. *YES, the Milky Way is a faint band of light across the sky that can be seen on a very dark night.* This is how the Milky Way appears with no city lights or clouds and no Moon. Ancient people who observed this band of light did not know what they were looking at, but they named it the Milky Way since it looked a little like a faint stream of spilled milk. In 1609 Galileo was the first to use a telescope to see that the faint band of light consists of countless stars.

B. *YES, the Milky Way is a spiral-shaped group of many stars.* By measuring the speed and location of stars and clouds of gas and dust, it has been possible to recognize the overall spiral shape of our galaxy. It is similar to many others that we can see from a significant distance.

C. *YES, the Milky Way is a galaxy.* There are billions of galaxies in the universe. Our Milky Way is one of them.

D. *YES, the Milky Way is our home in the universe.* Our home is in a tiny corner of the Milky Way, about two-thirds of the way from the center in one of the spiral arms.

E. *NO, the Milky Way is NOT a section of the universe smaller in size than our solar system.* It is much bigger than our solar system.

F. *NO, the Milky Way is NOT a milky-looking area that stretches across the entire universe.* It is much smaller than the entire universe.

G. *NO, the Milky Way is NOT a cloud of space dust and debris from planets and stars.* Although it does contain such debris, it also contains billions of stars.

H. *NO, the Milky Way is NOT a section of the universe that does not include our solar system.* Our solar system is a part of the Milky Way.

I. *YES, the Milky Way does contain billions of stars bound together by gravity.* There are billions of galaxies and each one, including the Milky Way, consists of a gravitationally bound cluster of billions of stars.

J. *NO, the Milky is NOT part of the Virgo cluster of galaxies, 54 million light-years from Earth.* Since we are inside the Milky Way galaxy, it cannot be far away from Earth.

In summary, the Milky Way is a galaxy—a large group of billions of stars. Since the solar system is inside the Milky Way, we cannot see it as a spiral. However, on a very dark and clear night we can see the disc edge-on as a faint band of light that stretches across the sky. If we could view it from far out in space, it would look like a lighted pinwheel with a bar across the middle and a bright center. Excellent time-lapse views of the Milky Way observed from Earth are available at: *www.youtube.com/watch?v=hRs8aseQikA&feature=related*

A video of an imaginary journey from Earth to outside the Milky Way is available at *www.youtube.com/watch?v=sm-ucbDVyRU.*

Administering the Probe

This probe can be used with students in middle school and high school. This justified list probe can also be used as a card sort in which the answer choices are put on cards and students sort them into the cards that describe the Milky Way and ones that do not (Keeley 2008). Be aware that for students to understand some of the answer choices in this probe, some prior instruction or experience in learning about galaxies is required. You may remove some of the answer choices that are unfamiliar to students.

Related Ideas in *Benchmarks for Science Literacy* (AAAS 2009)

6–8 The Universe

- The Sun is many thousands of times closer to the Earth than any other star. Light from the Sun takes a few minutes to reach the Earth, but light from the next nearest star takes a few years to arrive. The trip to that star would take the fastest rocket thousands of years.
- Some distant galaxies are so far away that their light takes several billion years to reach the Earth. People on Earth, therefore, see them as they were that long ago in the past.
- ★ The Sun is a medium-sized star located near the edge of a disc-shaped galaxy of stars, part of which can be seen as a glowing band of light that spans the sky on a very clear night.

★ Indicates a strong match between the ideas elicited by the probe and a national standard's learning goal.

Related Ideas in *National Science Education Standards* (NRC 1996)

9–12 The Origin and Evolution of the Universe

- Billions of galaxies, each of which is a gravitationally bound cluster of stars, now form most of the visible mass in the universe.

Related Research

- Friedman (2008) found that misconceptions regarding the size, extent, and composition of the Milky Way are rampant among middle school and high school students. Although no students participating in the study expressed the belief that the Milky Way is made of milk, they were at a loss regarding its composition aside from stars. Although some students were familiar with images of the Orion Nebula or the Ring Nebula, no one could identify the relationship of those objects to stars. Most students also considered the Sun, the Earth, and the solar system as unique members of the galaxy. The researcher also reported:

  Another prevailing misconception was that the Milky Way spans the entire Universe, or rather that there is no distinction between the two. Yet even among students who could correctly identify the Milky Way as a distinct entity residing in a greater universe, there was still a prevailing misconception regarding the relative size of the Solar System within the Milky Way. On the 60-foot scale of the model we constructed, students were shocked to learn that the entire Solar System could be contained within the width of a human hair. Most students expected the Solar System and galactic scales to be similar. Still, some could not identify that the Solar System resided within the Milky Way at all. (Friedman 2008, p. 179)

- Sadler et al. (2010) developed and validated a multiple-choice assessment item test bank for astronomy and space science in which the alternative answers for each question were based on research findings about common misconceptions. Results were reported for a sample of 7,599 students and their 88 teachers on the MOSART (Misconceptions-Oriented Standards-Based Assessment Resources for Teachers) website: *www.cfa.harvard.edu/smgphp/mosart/aboutmosart_2.html.* Items are listed for grades K–4, 5–8, and 9–12. Since each question has five answer choices, the odds of selecting the correct answer by chance are 20%. One item was as follows:

  Which answer shows the most accurate pattern of the three objects in order from closest object to the Earth to farthest from the Earth?

  A. Center of Milky Way → Andromeda galaxy → North Star
  B. Center of Milky Way → North Star→ Andromeda galaxy
  C. Andromeda galaxy → North Star → center of Milky Way
  D. North Star → Andromeda galaxy → center of Milky Way
  E. North Star → center of Milky Way → Andromeda galaxy

  The correct answer, E, was given by 36%. According to the researchers: "Although the correct choice was the most popular, 27% chose C, which places the North Star farther away than the Andromeda galaxy. Clearly those students (as well as the 35% combined choosing A, B or D) have little understanding of the relative distance of objects in the sky."

Suggestions for Instruction and Assessment

Although elementary students may have heard of the Milky Way, the concept of galaxies is not addressed until the middle school and high school levels. Helping students understand that our planet and solar system is inside the Milky Way should include lessons about galaxies in general, and about our galaxy in particular. Following are a few suggestions:

- There are many beautiful images of galaxies on the web; see, for example, *http://hubblesite.org/gallery/album/galaxy*. An open-ended activity in which students classify galaxies according to their overall shape can be fun, and can also give students a good overview of the many types of galaxies in the universe.
- A popular activity that can help students envision themselves living inside the Milky Way is to have students write their "galactic address." That is, ask the students to write their return address on a letter to an imaginary friend in another galaxy. They will start with their normal address, then expand it to include their country, continent, planet, planetary system, and galaxy (see "Activity H1. Your Galactic Address" in Fraknoi 2011).
- One of the most interesting stories in the history of science was the discovery that our Sun is one star among about 200 billion in a huge cluster of stars called the Milky Way galaxy, and that there are billions of galaxies in the universe. Various chapters in the story can be researched by teams of students using the internet. For example, "Everyday Cosmology" at *http://cosmology.carnegiescience.edu* includes the following facts:
    - In 1609 Galileo is the first to see that the Milky Way consists of stars.
    - In 1781 William Herschel realizes we probably live within a disc of stars.
    - In 1912 Henrietta Leavitt discovers how to find the distance to millions of stars.
    - In 1920 Harlow Shapley finds our place in the Milky Way.
    - In 1923 Edwin Hubble measures the distance to the Andromeda galaxy.
    - In 1929 Hubble discovers that galaxies are moving apart from each other.
- You might explain to the students that before Edwin Hubble measured the distance to the Andromeda galaxy, many astronomers believed that there was just one galaxy in space—our own Milky Way—and that the cloudlike clusters of stars that could be seen through telescopes were part of our own galaxy.
- High school students should learn as much as possible about the overall shape of the Milky Way and its constituents; the following websites have maps of the Milky Way:
    - *www.atlasoftheuniverse.com/milkyway.html*
    - *www.universetoday.com/22766/map-of-the-milky-way*

References

American Association for the Advancement of Science (AAAS). 2009. Benchmarks for science literacy online. *www.project2061.org/publications/bsl/online*

Fraknoi, A., ed. 2011. *Universe at your fingertips 2.0.* San Francisco: Astronomical Society of the Pacific. Available at *www.astrosociety.org/uayf/index.html*

Friedman, R. B. 2008. The Milky Way model. *Astronomy Education Review* 7 (2): 176–180. *http://aer.aas.org/resource/1/aerscz/v7/i2/p176_s1*

Keeley, P. 2008. *Science formative assessment: 75 practical strategies for linking assessment, instruction, and learning.* Thousand Oaks, CA: Corwin Press.

National Research Council (NRC). 1996. *National science education standards.* Washington, DC: National Academies Press.

Sadler, P. M., H. Coyle, J. L. Miller, N. Cook-Smith, M. Dussault, and R. R. Gould. 2010. The Astronomy and Space Science Concept Inventory: Development and validation of assessment instruments aligned with the K–12 National Science Standards. *Astronomy Education Review* 8 (1). *http://aer.aas.org/resource/1/aerscz/v8/i1/p010111_s1.* Questions and findings for each question are available at the MOSART website: *www.cfa.harvard.edu/smgphp/mosart*

Expanding Universe

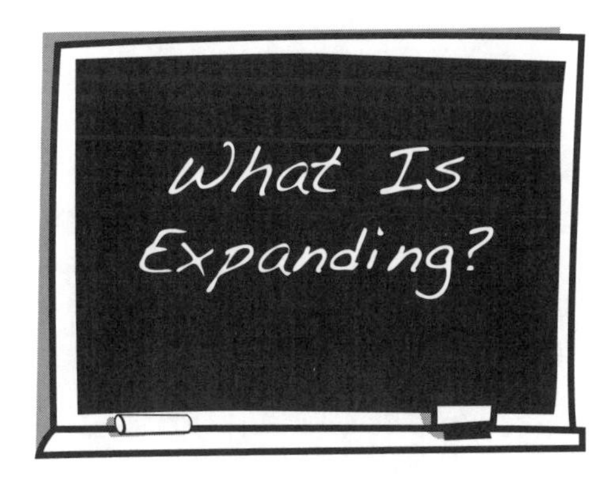

It has been more than 80 years since the astronomer Edwin Hubble discovered that the universe is expanding. He was able to determine that fact by observing the speed and direction that galaxies are moving. Nearly all galaxies are moving away from our galaxy, and the more distant galaxies are moving away faster. That means that all of the galaxies in the universe (or at least the material from which they were formed) were all together around 14 billion years ago, and they have been moving apart ever since. That is a scientific fact. But the question of *what* is expanding is part of the big bang theory. According to this theory, what, exactly, is expanding? Circle the answer that best matches your interpretation of the big bang theory.

A Matter is expanding into a huge empty void.

B Space is expanding or stretching, so the distance between galaxies is growing.

C Space and matter are expanding, so galaxies are getting bigger and moving apart.

Explain your thinking. Describe what you know about the big bang theory to support your answer.

____

____

____

____

Expanding Universe

Teacher Notes

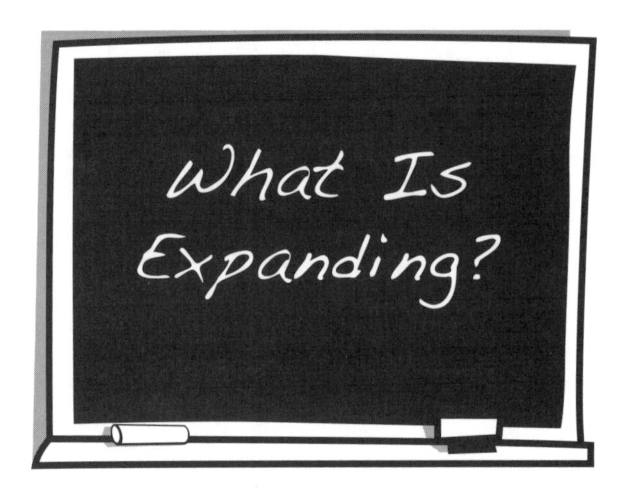

Purpose

The purpose of this assessment probe is to elicit students' ideas about the big bang theory. The probe is designed to reveal the extent of their qualitative understanding of what is expanding according to the theory.

Related Concepts

Big bang theory
Galaxies
Speed of light

Explanation

The best answer is B: Space is expanding or stretching, so the distance between galaxies is growing. Before the big bang neither space nor time nor matter as we know it existed. Now space itself is expanding, so the galaxies are seen to be moving farther apart. Strange as it may sound, there was no "space" in which the big bang exploded. In the instant of the big bang, space, time, energy, and matter came into being. Ever since, space itself has been expanding, or stretching. As a result, the galaxies have been moving apart from each other.

A common analogue is to think of a loaf of raisin bread, in which the raisins are sprinkled throughout the dough. As the bread expands the raisins move farther and farther apart. From the viewpoint of any one raisin it appears as though all the others are moving away, and the ones that are the farthest away are moving away the fastest. The raisins themselves (representing the galaxies) do not change size or shape—it is the dough (representing space) that is expanding.

Like all analogues, the raisin bread universe is not perfect. A loaf of bread expands inside an oven, whereas the universe makes its own space; and a loaf of bread has a center, whereas the universe has no center. Nonetheless, if you picture yourself inside the raisin bread, sitting on a raisin, the analogue is useful.

Administering the Probe

This probe is designed primarily for high school students, although it could be used with middle school students as an introduction to the big bang.

Related Ideas in *Benchmarks for Science Literacy* (AAAS 2009)

9–12 The Universe

- On the basis of scientific evidence, the universe is estimated to be over ten billion years old. The current theory is that its entire contents expanded explosively from a hot, dense, chaotic mass.
- Stars condensed by gravity out of clouds of molecules of the lightest elements until nuclear fusion of the light elements into heavier ones began to occur. Fusion released great amounts of energy over millions of years.
- Eventually, some stars exploded, producing clouds of heavy elements from which other stars and planets could later condense. The process of star formation and destruction continues.
- Increasingly sophisticated technology is used to learn about the universe. Visual, radio, and x-ray telescopes collect information from across the entire spectrum of electromagnetic waves; computers handle data and complicated computations to interpret them; space probes send back data and materials from remote parts of the solar system; and accelerators give subatomic particles energies that simulate conditions in the stars and in the early history of the universe before stars formed.
- Mathematical models and computer simulations are used in studying evidence from many sources in order to form a scientific account of the universe.

Related Ideas in *National Science Education Standards* (NRC 1996)

9–12 The Origin and Evolution of the Universe

- ★ The origin of the universe remains one of the greatest questions in science. The "big bang" theory places the origin between 10 and 20 billion years ago, when the universe began in a hot dense state; according to this theory, the universe has been expanding ever since.
- Early in the history of the universe, matter, primarily the light atoms of hydrogen and helium, clumped together by gravitational attraction to form countless trillions of stars. Billions of galaxies, each of which is a gravitationally bound cluster of billions of stars, now form most of the mass in the universe.

Related Research

- Sadler (1992) developed a written test to measure high school students' understanding of astronomy concepts. The test was administered to 1,414 students in grades 8–12 who were just starting an Earth science or astronomy course. One of the questions was as follows (percentage of students who chose each answer shown in parentheses):

  When the observable universe was half its present age, it was:
  A. larger than it is now (12%)
  B. smaller than it is now (39%)
  C. roughly the same size as it is now (23%)
  D. exactly the same size as it is now (9%)
  E. collapsed into a black hole (15%)
  The correct answer, B, was given by 39%. This is twice the number to be expected if students were guessing, so it suggests that

★ Indicates a strong match between the ideas elicited by the probe and a national standard's learning goal.

some students were aware that the universe is expanding.

- Lightman, Miller, and Leadbeater (1987) conducted a telephone survey of adults (age 18 and over). In response to the question "Do you think the universe is getting bigger in size, smaller in size, or remaining about the same size?" only 24% responded that the universe is increasing in size. The majority expressed the idea that the universe is static. Greater preference for an expanding universe was found among males, college graduates, those younger than age 50, and those who were not church members. This study went on to probe for the reasons supporting each individual's belief in the size of the universe. The most prevalent reason was "observation." The stars in the night sky appear motionless, and this observation appears to strongly motivate the belief in a static universe. Of the respondents, 19% expressed negative feelings about the discovery of an expanding universe. (Further details are provided in Lightman and Miller 1989.)
- The Lightman, Miller, and Leadbeater (1987) study also reported the results of a written questionnaire in which 83 high school students were asked, "If astronomers learned that the universe is increasing in size, with all the galaxies moving away from each other, how would this make you feel?" The researchers categorized the responses and found that 23% expressed positive responses, which included curiosity and a concern that space exploration might be more difficult if the universe were expanding. The researchers found that 43% expressed negative associations, including fear of unknown change, loss of control, a sense of helplessness, feelings of insignificance, and worries about possible destruction and death. The remaining responses were neither negative nor positive. In contrast to the religious associations in the adult survey, none of the high school students mentioned religion. The authors concluded that there appears to be a strong emotional component related to beliefs concerning the nature of the universe.
- In a series of interviews of Italian 11-year-olds, a variety of interesting explanations were given for a static universe. One student put it succinctly: "The stars do not move. If only one will, all the universe will be untidy" (Viglietta 1986, p. 369). All but the highest-performing students appear to prefer the idea that the universe is constant in size over the idea that it is expanding.

Suggestions for Instruction and Assessment

- This probe can be combined with "Is It a Theory?" in *Uncovering Student Ideas in Science, Vol. 3: Another 25 Formative Assessment Probes* (Keeley, Eberle, and Dorsey 2008) to establish what students think a theory is before they learn about the big bang theory.
- It is not appropriate to teach the big bang theory to elementary school students. Middle school students can be introduced to the essential ideas of the big bang theory: that galaxies are seen to be moving farther and farther apart, implying that they were once all together in the same place, billions of years ago.
- High school students should have an opportunity to learn essential concepts of the big bang theory. It is also important for high school students to learn about the evidence in support of the big bang, including at least:
    - Use of spectroscopic "red shift" or "blue shift" to measure the speed of a galaxy toward or away from us.

- The observation that more distant galaxies are moving away from us at a faster rate, known as Hubble's law.
- The interpretation of Hubble's law, using something like the raisin bread analogue, to explain why astronomers believe all galaxies were together in one place about 13.7 billion years ago.
- The concept that space, time, energy and matter were all created at that instant in time, now known as the big bang.
- Predictions of the big bang theory that 3°K background radiation would be observed coming from all directions.
- Detection of 3°K background radiation many years later, providing extremely strong evidence that big bang theory does indeed describe the universe at the beginning of time.
- The idea that, because the speed of light is constant, as we look at more distant galaxies, we are looking back in time, thus seeing how galaxies appeared when the universe was relatively young.

- Many educational resources are available about the big bang theory. See, for example, "Galaxies and the Universe" (astronomy background article 3.20) in *Universe at Your Fingertips 2.0* (Fraknoi 2011). A summary of key steps in the development of today's conception of the origin of the universe is found at *http://cosmology.carnegiescience.edu.*

References

American Association for the Advancement of Science (AAAS). 2009. Benchmarks for science literacy online. *www.project2061.org/publications/bsl/online*

Fraknoi, A., ed. 2011. *Universe at your fingertips 2.0.* San Francisco: Astronomical Society of the Pacific. Available at *www.astrosociety.org/uayf/index.html*

Keeley, P., F. Eberle, and C. Dorsey. 2008. *Uncovering student ideas in science, vol. 3: Another 25 formative assessment probes.* Arlington, VA: NSTA Press.

Lightman, A. P., J. D. Miller, and B. J. Leadbeater. 1987. Contemporary cosmological beliefs. In *Second International Seminar on Misconception and Educational Strategies in Science and Mathematics*, ed. J. D. Novak, pp. 309–321. Ithaca, NY: Cornell University Press.

Lightman, A. P., and J. D. Miller. 1989. Contemporary cosmological beliefs. *JSTOR: Social Studies of Science* (19): 127–136.

National Research Council (NRC). 1996. *National science education standards.* Washington, DC: National Academies Press.

Sadler, P. M. 1992. The initial knowledge state of high school astronomy students. Doctoral diss., Graduate School of Education, Harvard University.

Viglietta, M. L. 1986. Earth, sky and motion: Some questions to identify pupils' ideas. In Proceedings of the GIREP Conference 1986: Cosmos—An Educational Challenge, pp. 369–370. Copenhagen: European Space Agency.

Is the Big Bang "Just a Theory"?

Four college students—an astronomy student, a history student, a paleontology student, and a chemistry student—were discussing the big bang. This is what they said:

Astronomy student: "In astronomy the evidence for the big bang is that today all galaxies are racing apart from each other. I made some of the measurements myself! So we know that at one point in time the galaxies must have been together in one place at one time—that's the big bang."

History student: "No amount of evidence will convince me—if no one was there to record it, then we can never be sure. The big bang is just a theory."

Paleontology student: "No one was around when the dinosaurs lived, but we have evidence—fossils—that the dinosaurs existed. We have evidence of the big bang even though we weren't there when it happened."

Chemistry student: "To me the most important evidence that the big bang occurred is that when you look out into the universe today the most abundant element by far is hydrogen, followed by helium and a little lithium. That's exactly what you'd expect if hydrogen and helium were formed in a big bang, and the other elements were added later as stars formed and died."

Whose argument do you **least** accept? ______________________________

Explain your thinking on the back of this page. Evaluate each student's argument.

Is the Big Bang "Just a Theory"?

Teacher Notes

Purpose

The purpose of this assessment probe is to elicit students' ideas about the big bang as a theory. The probe is designed to find out if students can distinguish a *scientific theory* from the use of *theory* in everyday conversation.

Related Concepts

Big bang theory

Explanation

The history student's argument is the least acceptable. She thinks of the word *theory* as a guess, as in "I have a theory that the Red Sox will win the World Series this year." This student's approach to evidence is much like that of a lawyer or judge—if there was no eyewitness then we can't be sure what happened.

The three science students all understand the nature of a scientific theory as a framework for explaining numerous observations and a means for generating new predictions in order to better understand the natural world. A scientific theory is based on observations that have been tested and are widely supported by the scientific community. The science students are attempting to convince the history student of the validity of the big bang theory by demonstrating the strength of the evidence that supports it.

Administering the Probe

The probe is primarily designed for high school students and is best used to spark discussions about theories. Most students will be able to answer the first part of this probe. What is important is to examine the second part of the probe to learn more about your students' ideas about the explanatory nature of theories. Ask your students to consider the viewpoint of each speaker in turn. What is the speaker's point of view? What are the strengths and/or weaknesses in each argument? Ask the students to come up with their own rebuttal for or against the history student's statement.

Next focus on the use of the term *theory*. What are the two different meanings of the word in the probe? How did each speaker use the term *theory*? Might it be useful if we adopted different words to represent these two very different concepts?

One more follow-up question might focus on the words *believe* versus *understand*. Is the history student being asked to believe that the theory is true or just to understand the evidence that supports it? What is the difference between beliefs and understandings?

Related Ideas in *Benchmarks for Science Literacy* (AAAS 2009)

9–12 The Universe

- On the basis of scientific evidence, the universe is estimated to be over ten billion years old. The current theory is that its entire contents expanded explosively from a hot, dense, chaotic mass.
- Stars condensed by gravity out of clouds of molecules of the lightest elements until nuclear fusion of the light elements into heavier ones began to occur. Fusion released great amounts of energy over millions of years.
- Eventually, some stars exploded, producing clouds of heavy elements from which other stars and planets could later condense. The process of star formation and destruction continues.

9–12 The Scientific World View

- In science, the testing, revising, and occasional discarding of theories, new and old, never ends. This ongoing process leads to a better understanding of how things work in the world but not to absolute truth.

Related Ideas in *National Science Education Standards* (NRC 1996)

9–12 The Origin and Evolution of the Universe

- The origin of the universe remains one of the greatest questions in science. The "big bang" theory places the origin between 10 and 20 billion years ago, when the universe began in a hot dense state; according to this theory, the universe has been expanding ever since.

9–12 Nature of Scientific Knowledge

★ Science distinguishes itself from other ways of knowing and from other bodies of knowledge through the use of empirical standards, logical arguments, and skepticism, as scientists strive for the best possible explanations about the natural world.

Related Research

- Sadler et al. (2010) developed and validated a multiple-choice assessment item test bank for astronomy and space science in which the alternative answers for each question were based on research findings about common misconceptions. Results were reported for a sample of 7,599 students and their 88 teachers on the MOSART (Misconceptions-Oriented Standards-Based Assessment Resources for Teachers) website: *www.cfa.harvard.edu/smgphp/mosart/aboutmosart_2.html*. Items are listed for grades K–4, 5–8, and 9–12. Since each question has five answer choices, the odds of selecting the correct answer by chance are 20%. One item was as follows:

★ Indicates a strong match between the ideas elicited by the probe and a national standard's learning goal.

The "Big Bang" refers to the origin of:
A. the Sun
B. Earth
C. our solar system
D. the Milky Way galaxy
E. the universe
The correct answer, E, was given by 52%. According to the researchers: "Although the majority of students answered correctly, a quarter of students in our sample indicated that the Big Bang referred to the origin of the solar system (C). No other response was chosen by more than 10% of students."

- Shipman et al. (2002) investigated the responses of 340 undergraduate students enrolled in an introductory astronomy course that included discussions of views on the topic of religion and cosmic evolution. Data were gathered from the entire class and from in-depth interviews and focus groups with 19 of the students. There was little or no resistance to science and religion in a science class at a secular university, and approximately half of the students engaged with the issue. Some students with strong religious views that conflicted with the big bang theory were able to both understand the scientific theory and its evidentiary base, while others struggled for resolution or failed to resolve their conflicts.
- Brickhouse et al. (2000, 2002) expanded on the findings of the investigation by Shipman et al. (2002) by describing students' changing views of the nature of evidence in astronomy, and the nature of theory. With regard to the nature of theory, the instructor in the course defined a *theory* as an explanation that is broad in scope and fertile, in the sense that it not only provides satisfactory answers to old questions but also generates new ones. Data from student essays indicated that only about one in five students could provide adequate explanations for where theories come from and how they can be modified. Some students persisted in their misconceptions that hypotheses and theories are tentative, while facts and laws are absolute; as well as the idea that a theory can become a law when it is "proven" to be true. Further, the researchers found that how students talk and write about the nature of science depends on the particular scientific topic under discussion. For example, when discussing the theory of gravitation, many students refused to think of it as a theory, but rather thought of it as a fact. In explaining why, one student said, "Because how else would we be planted here? You know? Otherwise we'd be floating around and stuff. I think it's been proven. Why else would stuff fall towards the Earth and like with the planets?" (Brickhouse et al. 2002, p. 578). In contrast, the students thought very differently about the big bang theory. When asked to explain why they think it is indeed a theory, one student said, "Because it has to do with time and planets and stuff like that. But you know, it's not proven. No one was there when it happened" (Brickhouse et al. 2002, p. 583).
- Research shows that a deep, conceptual understanding of the nature of science has not been attained (Lederman et al. 2002), despite attempts to improve both students' and teachers' view of the nature of science.

Suggestions for Instruction and Assessment

- This probe can be preceded with "Is It a Theory?" from *Uncovering Student Ideas in Science, Vol. 3: Another 25 Formative Assessment Probes* (Keeley, Eberle, and Dorsey 2008) to further elicit students' ideas about scientific theories.

- Virtually every culture, from the earliest prehistory to the present, has passed down from generation to generation stories about the origin of the universe. These stories have had important cultural value as they offered explanations for eternal human questions about where we come from and what our future may be. Core ideas about how our universe came to be are part of our modern cultural heritage that all students should have an opportunity to learn. However, it is important that students not be taught about the big bang as simple fact, but rather as a theory that gradually emerged as evidence accumulated over the past few decades. Consequently, it is important that students learn about the nature of scientific theories, so that the big bang theory of the universe can be viewed in its proper scientific context.
- Discussion of scientific theories and the notion of evidence can begin in the earliest grades. When students make statements, ask them to tell you why they believe that to be the case. Encourage students to discuss ideas, to sometimes disagree and have respectful discussions about why they disagree. However, the theory and nature of evidence of the big bang theory is too advanced to be discussed at the elementary school level.
- Middle school students should have opportunities to further develop their abilities to communicate their ideas, and to critically and respectfully discuss ideas expressed by their peers. The early rivalry between the big bang theory and the steady-state theory for the origin of the universe can be presented at an appropriate level, as well as a qualitative discussion for the evidence in support of the big bang.
- An introductory high school physics course will provide the foundation that students need to understand the kinds of "red shift" measurements that led to the development of Hubble's law. Students should also have sufficient logical reasoning skills to understand how Hubble's law led to the conclusion that all galaxies were once together in the same place and led to an estimate for the age of the universe.
- Although the mathematical prediction of 3°K blackbody radiation will be well beyond most high school students' capabilities, it is important that they learn about that prediction—attributed primarily to Ralph Alpher—and the eventual discovery of the radiation from all directions that led to the rejection of the steady-state theory in favor of the big bang theory. The many interesting human stories involved can be found at *http://cosmology.carnegiescience.edu*.
- At the conclusion of a unit on the big bang theory of the universe, it may be instructive for you and your students to revisit Probe 44, "Expanding Universe." Look for evidence that your students have a solid understanding of what scientific theories are all about. The important point is not that they "believe" in the big bang theory but rather that they understand why the theory was developed, what the theory says, and the evidence that supports it.

References

American Association for the Advancement of Science (AAAS). 2009. Benchmarks for science literacy online. *www.project2061.org/publications/bsl/online*

Brickhouse, N. W., Z. R. Dagher, W. J. Letts IV, and H. L. Shipman. 2000. Diversity of students' views about evidence, theory, and the interface between science and religion in an astronomy course. *Journal of Research in Science Teaching* 37 (4): 340–362.

Brickhouse, N. W., Z. R. Dagher, H. L. Shipman, and W. J. Letts IV. 2002. Evidence and war-

rants for belief in a college astronomy course. *Science and Education* 11: 573–588.

Keeley, P., F. Eberle, and C. Dorsey. 2008. *Uncovering student ideas in science, vol. 3: Another 25 formative assessment probes.* Arlington, VA: NSTA Press.

Lederman, N., F. Abd-El-Khalick, R. Bell, and R. Schwartz. 2002. Views of nature of science questionnaire: Toward valid and meaningful assessment of learner's conceptions of nature of science. *Journal of Research in Science Teaching* 39 (6): 497–521.

National Research Council (NRC). 1996. *National science education standards.* Washington, DC: National Academies Press.

Sadler, P. M., H. Coyle, J. L. Miller, N. Cook-Smith, M. Dussault, and R. R. Gould. 2010. The Astronomy and Space Science Concept Inventory: Development and validation of assessment instruments aligned with the K–12 National Science Standards. *Astronomy Education Review* 8 (1). *http://aer.aas.org/resource/1/aerscz/v8/i1/p010111_s1.* Questions and findings for each question are available at the MOSART website: *www.cfa.harvard.edu/smgphp/mosart*

Shipman, H. L., N. W. Brickhouse, Z. Dagher, and W. J. Letts. 2002. Changes in student views of religion and science in a college astronomy course. *Science Education* 86 (4): 526–547.

Index

Note: Page numbers in *italics* refer to charts.

Index

Index

Index

Index

Index

Index

Index

Index